Stephan W. Anzengruber

The discrepancy principle for Tikhonov regularization in Banach spaces

Stephan W. Anzengruber

The discrepancy principle for Tikhonov regularization in Banach spaces

Regularization properties and rates of convergence

Südwestdeutscher Verlag für Hochschulschriften

Impressum/Imprint (nur für Deutschland/only for Germany)
Bibliografische Information der Deutschen Nationalbibliothek: Die Deutsche Nationalbibliothek verzeichnet diese Publikation in der Deutschen Nationalbibliografie; detaillierte bibliografische Daten sind im Internet über http://dnb.d-nb.de abrufbar.
Alle in diesem Buch genannten Marken und Produktnamen unterliegen warenzeichen-, marken- oder patentrechtlichem Schutz bzw. sind Warenzeichen oder eingetragene Warenzeichen der jeweiligen Inhaber. Die Wiedergabe von Marken, Produktnamen, Gebrauchsnamen, Handelsnamen, Warenbezeichnungen u.s.w. in diesem Werk berechtigt auch ohne besondere Kennzeichnung nicht zu der Annahme, dass solche Namen im Sinne der Warenzeichen- und Markenschutzgesetzgebung als frei zu betrachten wären und daher von jedermann benutzt werden dürften.

Coverbild: www.ingimage.com

Verlag: Südwestdeutscher Verlag für Hochschulschriften GmbH & Co. KG
Heinrich-Böcking-Str. 6-8, 66121 Saarbrücken, Deutschland
Telefon +49 681 37 20 271-1, Telefax +49 681 37 20 271-0
Email: info@svh-verlag.de

Approved by: Linz, JKU, Diss., 2011

Herstellung in Deutschland (siehe letzte Seite)
ISBN: 978-3-8381-3135-1

Imprint (only for USA, GB)
Bibliographic information published by the Deutsche Nationalbibliothek: The Deutsche Nationalbibliothek lists this publication in the Deutsche Nationalbibliografie; detailed bibliographic data are available in the Internet at http://dnb.d-nb.de.
Any brand names and product names mentioned in this book are subject to trademark, brand or patent protection and are trademarks or registered trademarks of their respective holders. The use of brand names, product names, common names, trade names, product descriptions etc. even without a particular marking in this works is in no way to be construed to mean that such names may be regarded as unrestricted in respect of trademark and brand protection legislation and could thus be used by anyone.

Cover image: www.ingimage.com

Publisher: Südwestdeutscher Verlag für Hochschulschriften GmbH & Co. KG
Heinrich-Böcking-Str. 6-8, 66121 Saarbrücken, Germany
Phone +49 681 37 20 271-1, Fax +49 681 37 20 271-0
Email: info@svh-verlag.de

Printed in the U.S.A.
Printed in the U.K. by (see last page)
ISBN: 978-3-8381-3135-1

Copyright © 2012 by the author and Südwestdeutscher Verlag für Hochschulschriften GmbH & Co. KG and licensors
All rights reserved. Saarbrücken 2012

Abstract

In this work, we study Tikhonov regularization with general convex penalty terms Ψ for nonlinear inverse problems in Banach spaces. We are concerned with the case that only noisy data y^δ corresponding to a noise bound δ is at hand and the regularization parameter α is chosen according to an a-posteriori parameter choice rule known as Morozov's discrepancy principle.

Sufficient conditions on the operator and the noisy data for the existence of such a regularization parameter α are formulated and it is shown that minimizing the Tikhonov functional constitutes a regularization method when using the discrepancy principle to choose α. The regularized solutions converge to the Ψ-minimizing solutions in different topologies depending on properties of Ψ and the parameter choice rule under consideration satisfies the asymptotic relations $\alpha \to 0$, $\delta^q/\alpha \to 0$ as the noise level δ tends to 0.

We derive convergence rates using variational inequalities to generalize classical source and nonlinearity conditions. Rates are obtained first with respect to the Bregman distance and a Taylor-type distance and those results are combined to derive rates in norm or the penalty term topology.

For the special case of the sparsity promoting weighted ℓ_p-norms as penalty terms and for a searched-for solution, which is known to be sparse, the above results are applied and shown to yield convergence rates of up to linear order.

Finally, we study the autoconvolution operator and present numerical results.

Acknowledgment

I would like to thank my supervisor, Ronny R., who has not only introduced me to the topics of this thesis and given me the opportunity to work in this interesting field of mathematics, but has truly guided me throughout these last years with his experience and patience.

Without the support of my loving family this thesis would never have been possible. I, especially, thank my wife, Aneta, for always believing in me and what I do, my mum for giving me the opportunity to pursue my studies, as well as my grandma and my brothers.

Many thanks go to my colleagues in Linz for valuable discussions over lunch or coffee and countless helpful remarks: Stefan K., Andreas N., Mascha Z., Ismael B., Esther K., Jenny N. and Daniel W.

I also thank Patti L., who has been a great host during my stimulating research stay at Michigan State University, and Bernd H. for raising my interest in variational inequalities and sublevelsets and for co-refereeing this thesis.

Contents

1 Topological preliminaries   **7**
   1.1 The weak and weak-* topologies 7
   1.2 Properties of functionals and operators 10

2 Tikhonov regularization   **15**
   2.1 Formulation of the Tikhonov method 16
   2.2 ℓ_p-norms of frame coefficients in Hilbert spaces 22
   2.3 Classical Tikhonov regularization in Banach spaces 25
   2.4 Besov spaces 25
   2.5 TV regularization 27

3 Regularization properties   **31**
   3.1 Existence and Stability 31
   3.2 Convergence 33

4 The discrepancy functionals   **39**
   4.1 Monotonicity 41
   4.2 Continuity 43
   4.3 Asymptotic behaviour 48
   4.4 Strict monotonicity 49
   4.5 Subdifferentiability 51

5 The Discrepancy Principle   **55**
   5.1 Existence of α according to MDP 55
   5.2 Asymptotics 58
   5.3 Other formulations 62
   5.4 Computational aspects 63
   5.5 An academic example 66

6 Convergence rates   **75**
   6.1 Sublevelsets of $J_{\alpha,y}(x)$ and $\Psi(x)$ 76
      6.1.1 Sublevelsets of $\Psi(x)$ 76
      6.1.2 Sublevelsets of $J_{\alpha,y}(x)$ 77
      6.1.3 Relation between the sublevelsets of $\Psi(x)$ and $J_{\alpha,y}(x)$ 79

6.2	Bregman and Taylor distance	81
6.3	Variational inequalities as source and nonlinearity conditions on sublevelsets	82
6.4	Convergence rates in Bregman and Taylor distance	85
6.5	Convergence rates in norm and the penalty term topology	86
6.6	Local variational inequalities	89
6.7	Strict variational inequalities	91
6.8	Convergence rates under a scaling invariant nonlinearity condition	95
6.9	The linear case	97
6.10	Relation of variational inequalities to source and nonlinearity conditions	98

7 Sparse recovery 101
7.1 Assumptions and problem formulation 101
7.2 Variational Inequalities 102
7.3 Convergence rates 105

8 Autoconvolution 107
8.1 Properties 107
8.2 Formulation in wavelet coefficients 108
8.3 Numerical results 110

Introduction

In the last decade, *sparse recovery* has received considerable attention in the context of *inverse problems* as well as in various other fields of applied mathematics, such as compressed sensing, mathematical imaging and optimal control. Originally, finding sparse solutions of operator equations was relevant to signal (image) denoising and data compression, but the number of applications is constantly increasing and includes the design of optical layers, medical and astronomical imaging as well as economic market simulations and even the Netflix problem.

In order to obtain sparse solutions, one often times minimizes a Tikhonov-type variational functional with penalty terms that are not covered by the classical convergence theory (as laid out, for example, in [17]), namely, the sparsity-promoting ℓ_p-norms, with $1 \leq p < 2$, of coefficients with respect to a suitably chosen basis or frame of the underlying space. Especially the limiting case $p = 1$ is of great interest for two main reasons:

- Minimizing the ℓ_1-penalized least squares functional is the convex relaxation of the underlying NP-hard problem of finding the *sparsest* solution of a given operator equation, i.e., the solution with the smallest number of non-zero coefficients.

- In mathematical imaging the ℓ_1-norm of coefficients with respect to a suitably chosen wavelet basis approximates the *total variation* (TV) of an image. TV regularization will be discussed in more detail in Section 2.5.

With the development of efficient algorithms to solve the corresponding optimization problem [4, 7, 8, 13, 14, 15, 27, 44, 48, 50, 60], ℓ_p-variational regularization has become widely-used in practical applications and a thorough analysis of convergence properties of the underlying method is required. Recent publications in this direction include [1, 2, 6, 5, 9, 13, 18, 32, 33, 35, 24, 25, 56]. The present thesis contributes to this active field of mathematical research in as much as it provides results on regularization properties, convergence and the speed of convergence of convex, variational regularization methods in Banach spaces, which includes sparsity enforcing methods as well as total variation regularization, for a certain parameter choice rule, which is known as *Morozov's discrepancy principle*.

If we are interested in finding a quantity x from observed data y, which are connected through a (non-linear) operator equation of the form

$$F(x) = y, \qquad (1)$$

then noise in the data y may lead to large errors in the reconstruction, especially if the operator under consideration is ill-posed. More precisely, problem (1) is called *well-posed* in the sense of Hadamard [28, 29] if the following conditions are satisfied.

(i) For any y a solution x of (1) exists.

(ii) For any y the solution is unique.

(iii) The solution depends continuously on the data y.

Otherwise, if any of the above statemtents does not hold, then the problem is called *ill-posed*. Examples of physical and economical applications where ill-posed operator equations occur, include, but are by no means limited to, medical and astronomical imaging, inverse scattering and mathematical finance. Frequently the data y are corrupted by noise, for instance, if they were obtained through a measurement process which is subject to inaccuracy. We will indicate the noisy version of the data by y^δ. In this case, even if $y \in \text{rg}(F)$ admits a solution, nonetheless a solution of (1) with y replaced by y^δ may no longer exist.

If the data y, y^δ belong to a normed space Y, we may quantify the amount of noise in the data using a *noise level* δ, i.e., $\|y - y^\delta\| \leq \delta$. Then, as a first step towards making problem (1) mathematically more tangible, one can consider instead the minimization of the least-squares functional

$$J(x) = \left\| F(x) - y^\delta \right\|_Y^2. \qquad (2)$$

This formulation allows for the definition of a *least-squares solution* even if the noisy data do not belong to the range of the operator, $y^\delta \notin \text{rg}(F)$, provided that a minimizer exists. Such solutions are of special interest if the underlying space Y is a Hilbert space. In other spaces different powers of the norm as well as other penalty terms (e.g., the Kullback-Leibler divergence) may also be used. But problem (2) remains ill-posed in the sense that due to the non-linearity of F the solution may not be unique and may not depend continuously on the data.

One approach to overcome these difficulties is to replace (2) by a family $\{J_\alpha\}_{\alpha>0}$ of neighboring well-posed (or at least stable) problems, which incorporate additional a-priori knowledge of properties of the searched-for solution through a regularizing functional $\Psi(x)$, also called the *penalty term*. For the purpose of this work we will only be concerned with convex Ψ, results on non-convex regularization can be found, e.g., in [22, 23, 65]. Among all solutions x of (2) we are looking for the Ψ-*minimizing solutions*, i.e., the ones with the minimal value with respect to Ψ. As we will see, they can be approximated by the minimizers – denoted by x_α^δ – of the family of variational functionals

$$J_{\alpha,y^\delta}(x) = \left\| F(x) - y^\delta \right\|_Y^2 + \alpha \Psi(x), \qquad \alpha > 0. \tag{3}$$

The choice $\Psi(x) = \|x\|_X^2$, for x belonging to some Hilbert space X, constitutes the classical Tikhonov regularization. We refer the reader to [17] for further details in this respect.

We will repeatedly encounter the sparsity promoting weighted ℓ_p-norms with respect to a given basis or frame $\{\phi_\lambda\}_{\lambda \in \Lambda}$ of a Hilbert space X,

$$\Psi_{w,p}(x) = \sum_{\lambda \in \Lambda} w_\lambda \, |\langle \phi_\lambda, x \rangle|^p, \qquad 0 < w_0 \leq w_\lambda, 1 \leq p \leq 2. \tag{4}$$

Sparse representations of solutions are of strong interest, for example, in signal compression and astronomical imaging, where objects of interest like images are sparse, but their standard reconstructions are not. Enforcing sparsity adds knowledge on the solution and therefore improves the reconstruction.

Another important regularization term which falls into the scope of this survey is the total variation (TV) of a function x in the space of bounded variation $BV(\Omega)$,

$$TV(x) = \int_\Omega |Dx| \, dt,$$

where Dx denotes the distributional derivative of x. TV regularization is well established in mathematical imaging due to its good reconstruction of edges in cartoon-like images. Anistropic variants of TV have also been found to favour corners or rectangular structures.

The choice of the regularization parameter α in (3) turns out to be of crucial importance for the quality of the resulting reconstructions. Many different parameter choice rules (p.c.r.) have been proposed in the literature. They can roughly be divided into three categories.

A-priori. The regularization parameter is chosen depending only on the noise level, $\alpha = \alpha(\delta)$. The most common choice is $\alpha(\delta) = c \cdot \delta^r$ for some $0 < r < 2$ and $c > 0$. The main advantage of this class of p.c.r. is its simple implementation. The choice of the constant c typically does not influence the convergence properties of this method. It does, however, have a great impact on the reconstructions in practical applications. Moreover, in some cases convergence rate results for a-priori p.c.r. are proven under the additional assumption that the choice of r involves knowledge of certain properties of the searched-for solution $x^\dagger$, such as its smoothness. In most practical applications such knowledge will not be at hand.

A-posteriori. The regularization parameter is chosen depending on the noise level and on the noisy data, $\alpha = \alpha(\delta, y^\delta)$. Here, the main advantage is that the given noisy data have a direct impact on the choice of the regularization parameter. This comes at the cost of a higher computational effort in the determination of such a parameter. The most prominent representative of this class of p.c.r. is Morozov's discrepancy principle (MDP), which we study in the following chapters. When using MDP we choose $\alpha = \alpha(\delta, y^\delta)$ such that for some minimizer x_α^δ of (3)

$$\tau_1 \delta \leq \left\| F\left(x_{\alpha(\delta,y^\delta)}^\delta\right) - y^\delta \right\| \leq \tau_2 \delta, \qquad 1 \leq \tau_1 \leq \tau_2 \tag{5}$$

holds. This way of choosing the regularization parameter has been studied in great detail in its present formulation [1, 2, 6, 47] as well as in several related variations [17, 25, 38, 45, 54, 61, 62]. Other a-posteriori p.c.r. have been studied, for example, in [19, 55, 58, 59].

Noise-free or Heuristic. The regularization parameter is chosen depending only on the noisy data, $\alpha = \alpha(y^\delta)$. Some of the well-known examples of noise-free p.c.r. are general cross validation [20], the L-curve method [30, 31] and the quasi-optimality criterion [39]. A main drawback of these methods is known as Bakushinskii's veto [3], which states that Tikhonov regularization cannot be a convergent regularization method for noise-free p.c.r. However, under (reasonable) additional assumptions on the noise in the data, convergence results were shown, e.g., in [39] for the quasi-optimality criterion.

For a-priori parameter choice rules convergence and rates of convergence, also known as error estimates, of regularized solutions to a true solution for general convex penalty terms Ψ have been shown with respect to the Bregman distance for linear operators in [9, 51] and for the case of a nonlinear operator in [52]. For the functionals $\Psi_{w,p}$ in (4) these results were used to obtain convergence rates in norm (cf. [24, 42, 49]).

In [17] it has been shown that for linear operator equations and certain classes of regularization methods defined via spectral decomposition in Hilbert spaces, the discrepancy principle gives order optimal convergence rates. These results cover the classical Tikhonov regularization mentioned above, but not variational regularization methods with general convex penalty terms as in (3), and in particular not the functionals $\Psi_{p,w}$ in (4) for $p < 2$.

For the special case of denoising, where the operator under consideration is the identity in $L^2(\mathbb{R}^d)$ and an L^1-penalty term was chosen, optimal order convergence rate results were obtained with respect to the L^1-norm in [38]. It was also shown that the resulting regularization method does not saturate, in which case the discrepancy principle yields the same convergence rates as a-priori parameter choice rules.

For Tikhonov-type regularization with linear operator and general convex penalty terms as in (3) with regularization parameter chosen according to MDP, Bonesky [6] showed convergence rate results with respect to the Bregman distance in reflexive Banach spaces and his results were generalized to non-linear operators in [1] adding an additional condition on the structure of the non-linearity in F.

Finally, the residual method was studied in the report [25]. It is closely related to the discrepancy priniple and in r-convex Banach spaces, $r \geq 2$, convergence rates in norm were derived when using the penalty term $\Psi(x) = \|x\|_X^r / r$. Moreover, for linear operators additional convergence rates were provided under the assumption that the unknown solution is sparse.

Once the regularization parameter has been chosen, it remains to compute the related regularized solution as the minimizer of the Tikhonov-type functional (3). Different methods to achieve this can be found in [8, 13, 27, 48, 50].

In this survey, we study Morozov's discrepancy principle (MDP) (5) as the p.c.r. for Tikhonov-type variational regularization of non-linear ill-posed problems (1) in Banach spaces.

In the first chapter we summarize topological and functional analytical results which are essential to the techniques employed in our proofs, then we formally introduce Tikhonov regularization and formulate our technical assumptions in Chapter 2. Examples of penalty terms that are covered by our theory are powers of the norm in Hilbert and Banach spaces, weighted ℓ_p-norms of frame coefficients in Hilbert spaces and wavelet coefficients in Besov spaces, as well as total variation regularization in the space of bounded variation.

We shall see that the regularized problems (3) are well-posed in Chapter 3 and that if the parameter α is chosen according to MDP, then the corresponding regularized solutions x_α^δ converge (weakly-*) to a Ψ-minimizing solution $x^\dagger$. For penalty terms fulfilling a generalized Kadec property we show convergence in norm or even with respect to the penalizing functional Ψ, i.e., that

$$\Psi(x_\alpha^\delta - x^\dagger) \to 0 \quad \text{as} \quad \delta \to 0. \tag{6}$$

To study the question of existence of a parameter fulfilling MDP, in Chapter 4 we collect properties of the Tikhonov functional evaluated at its minimizer, $J_{\alpha,y^\delta}(x_\alpha^\delta)$, for fixed y^δ in dependence of α and of two closely related functionals, which together we will refer to as discrepancy functionals.

Then – using the properties of the discrepancy functionals – we derive a sufficient condition for the existence of a regularization parameter $\alpha = \alpha(\delta, y^\delta)$ fulfilling the discrepancy principle in Chapter 5. As in the well studied case of classical Tikhonov regularization, we will be able to show that standard conditions on the operator F suffice to guarantee the existence of a positive regularization parameter. Some regularization results hold for arbitrary parameter choice rules which satisfy the asymptotical relations

$$\alpha(\delta, y^\delta) \to 0 \quad \text{and} \quad \delta^q/\alpha(\delta, y^\delta) \to 0$$

as $\delta \to 0$ and we also show that the discrepancy principle belongs to this class of p.c.r. Moreover, we discuss other formulations of MDP available in the literature as well as numerical aspects of computing a parameter α according to MDP and study an instructive, one-dimensional example where – depending on the noise level δ and the noisy data y^δ – MDP has infinitely many, one or no solutions.

In Chapter 6 we derive the main results regarding rates of convergence of the Tikhonov regularized solutions to a Ψ-minimizing solution. The key ingredient for our analytic results are variational inequalities (VIE), which generalize classical source- and nonlinearity conditions. We will see that the domain on which the VIE are required to hold may be restricted compared to a-priori parameter choice rules when working with MDP, which leads to a significant simplification in the formulation of the VIE. Rates are first derived with respect to the Bregman distance and a Taylor-type distance, and then these results are combined to obtain rates in norm or even with respect to the penalty term.

In Chapter 7 we consider the example of sparse recovery in Hilbert spaces, which we will find to be a special case of the framework described in the earlier parts, and we show that if the searched-for solution $x^\dagger$ is sparse, then convergence rates of up to linear order can be observed when penalizing with $\Psi_{w,1}$ as defined in (4) for non-linear operators using MDP in combination with variational inequalities.

Finally, we will present the autoconvolution operator as a numerical example in Chapter 8, where the theoretically established results are verified, when reconstructing sparse solutions with respect to the Haar wavelet basis and using an ℓ_1 penalty of the wavelet coefficients.

Chapter 1

Topological preliminaries

In this chapter we collect the main definitions and analytical properties of Banach spaces that will be of consequence to us. The proofs of (some of) these results require deep insights from topology and functional analysis which go beyond the scope of this work. The interested reader is referred to [62, 63].

1.1 The weak and weak-* topologies

Definition 1.1. Let X be a Banach space over $\mathbb{R}$ with norm $\|.\|$. The dual space X^* of X is defined as

$$X^* = \{\varphi : X \to \mathbb{R} \mid \varphi \text{ linear, continuous }\}$$

with norm

$$\|\varphi\|_{X^*} = \sup\{ |\varphi(x)| \mid x \in X : \|x\| \leq 1\}.$$

For elements in the dual space we use the functional notation $\varphi(x)$ alongside the duality product notation $\langle \varphi, x \rangle_{X^*, X}$ as we see fit.

Proposition 1.2. *The dual space of any Banach space over $\mathbb{R}$ is again a Banach space over $\mathbb{R}$.*

Throughout this survey we will in general indicate all norms on Banach spaces simply by $\|.\|$ if it is clear from the context which space is referred to. Nevertheless, if we wish to emphasize the underlying space a subscript will be used, such as $\|.\|_{X^*}$, $\|.\|_X$, etc. Also, in what follows we will only be concerned with Banach spaces over $\mathbb{R}$. Therefore we will omit the reference to the underlying scalar field from here on.

Definition 1.3. On a Banach space X the topology $\mathcal{T}_w$ that is induced by the family of seminorms $\{p_\varphi : X \to \mathbb{R}_0^+ \mid \varphi \in X^*\}$, with

$$p_\varphi(x) = |\varphi(x)|$$

is called the *weak topology*. $(X, \mathcal{T}_w)$ is a locally convex Hausdorff space.

A local basis of a point $x_0 \in X$ in the weak topology is given by neighbourhoods of the form

$$U_{\Phi,\varepsilon}(x_0) = \{x \in X \mid \forall \varphi \in \Phi : p_\varphi(x - x_0) < \varepsilon\},$$

for any finite subset $\Phi \subset X^*$ and any $\varepsilon > 0$. The weak topology is Hausdorff as a consequence of the Hahn-Banach theorem. We now characterize the notion of convergence in $\mathcal{T}_w$.

Definition 1.4. A sequence $\{x_n\} \in X^\mathbb{N}$ is said to *converge weakly* (i.e., with respect to the weak topology) to an element $x \in X$, we write $x_n \rightharpoonup x$, if for all $\varphi \in X^*$

$$\varphi(x_n) \to \varphi(x) \text{ as } n \to \infty.$$

In the dual space X^* the weak topology is induced by the linear, continuous functionals in the *bidual space* X^{**} of X. Every Banach space X can be identified with a closed subspace of its bidual X^{**} due to the following proposition.

Proposition 1.5. *The mapping* $i_X : X \to X^{**}$, $(i_X(x))(\varphi) = \varphi(x)$ *is linear, continuous and injective with*

$$\|i_X(x)\|_{X^{**}} = \|x\|.$$

Therefore we can define the weak-* topology on X^* which is even weaker than the weak topology.

Definition 1.6. Let X be a Banach space and i_X as defined in Proposition 1.5. The topology $\mathcal{T}_{w^*}$ on X^* induced by the family of seminorms

$$\left\{ |i_X(x)| : X^* \to \mathbb{R}_0^+ \mid x \in X \right\}$$

is called the *weak-* topology*. $(X^*, \mathcal{T}_{w^*})$ is a locally convex Hausdorff space.

The corresponding notion of convergence is the weak-* convergence.

Definition 1.7. A sequence $\{\varphi_n\} \in (X^*)^\mathbb{N}$ is said to *converge weakly-** to an element $\varphi \in X^*$, we write $\varphi_n \rightharpoonup^* \varphi$, if

$$\varphi_n(x) \to \varphi(x) \text{ as } n \to \infty$$

for all $x \in X$.

An important fact about the weak-* topology is the following result due to Banach-Alaoglu.

Proposition 1.8. *The closed unit ball in X^* is weak-* compact. Therefore every bounded sequence in X^* contains a weak-* convergent subsequence.*

Clearly, every convergent sequence in X is also weakly convergent and every weakly convergent sequence in X^* is weak-* convergent and their respective limit points coincide. The converse statements are, however, not true.

Definition 1.9. A Banach space X is called *reflexive* if the mapping i_X in Proposition 1.5 is surjective.

Consequently, for every reflexive Banach space X it holds that $X \cong X^{**}$ but this condition is not sufficient (cf. [41, p. 25]). Hence, if X is reflexive it can be naturally endowed with the weak-* topology that carries over from X^{**}. The dual space X^* then generates both the weak and the weak-* topology on X which motivates the following result.

Proposition 1.10. *If X is reflexive then the weak topology and the weak-* topology on X coincide.*

Therefore the precompactness of bounded sets from Proposition 1.8 carries over to the weak topology in this case. For the following converse result reflexivity is not needed. It is a consequence of the theorem of Banach-Steinhaus.

Proposition 1.11. *Every weakly convergent sequence in a Banach space X is bounded.*

Let us define what we mean by convergence of a sequence to a set in any topology.

Definition 1.12. Let $(X, \mathcal{T})$ be a locally convex Hausdorff space. We say that a sequence $\{x_n\} \in X^{\mathbb{N}}$ converges to a set $A \subset X$ with respect to $\mathcal{T}$ if to every neighbourhood T of A in $\mathcal{T}$ there exists $N \in \mathbb{N}$ such that $x_n \in T$ for all $n \geq N$.

If the topology $\mathcal{T}$ is metrizable and $d : X \times X \to \mathbb{R}_0^+$ is the corresponding metric, then Definition 1.12 coincides with the classical notion of set valued convergence where a sequence $\{x_n\}$ in X converges to $A \subset X$ if

$$d(x_n, A) \to 0 \text{ as } n \to \infty.$$

and the distance of a point $x \in X$ to a set A with respect to d is given by

$$d(x, A) = \inf_{z \in A} d(x, z).$$

Sequences that converge to a set have some important properties that are of interest when studying the convergence of regularization methods for inverse problems.

Proposition 1.13. *If $x_n \to A$ w.r.t. $\mathcal{T}$ then the following statements hold as well:*

(i) *Each subsequence of $\{x_n\}$ has a subsequences that converges to some $a \in A$ w.r.t. $\mathcal{T}$.*

(ii) *The limit point of every convergent subsequence of $\{x_n\}$ lies in A.*

(iii) *If A consists of only one element a, then $x_n \to a$ w.r.t. $\mathcal{T}$.*

We conclude this section with the definition of weak sequential closedness.

Definition 1.14. A set $A \subset X$ of a Banach space X is *weakly (weak-*) sequentially closed* if it contains the limit points of every weakly (weak-*) convergent sequence $\{x_n\} \in A^{\mathbb{N}}$.

1.2 Properties of functionals and operators

We now discuss important properties of functionals and operators on topological spaces.

Definition 1.15. Let (X, τ_X) and (Y, τ_Y) be topological spaces. An operator $F : \operatorname{dom}(F) \subset X \to Y$ is called *sequentially closed* between τ_X and τ_Y, if for any sequence $\{x_n\} \subset \operatorname{dom}(F)$ such that

$$x_n \xrightarrow{\tau_X} x \quad \text{and} \quad F(x_n) \xrightarrow{\tau_Y} y$$

it holds that $x \in \operatorname{dom}(F)$ and $y = F(x)$.

At this point we introduce our notation regarding derivatives.

Definition 1.16. Let X, Y be locally convex spaces and let $\operatorname{dom}(F) \subset X$ be open. An operator $F : \operatorname{dom}(F) \to Y$ is differentiable at $x \in \operatorname{dom}(F)$ in direction $h \in X$ if the limit

$$dF(x;h) = \lim_{t \to 0} \frac{F(x+th) - F(x)}{t} = \frac{d}{dt} F(x+th) \bigg|_{t=0}$$

exists. $dF(x; h)$ is called the *directional derivative* at x in direction h.

Definition 1.17. If an operator $F : \operatorname{dom}(F) \to Y$ is differentiable at $x \in \operatorname{dom}(F)$ in every direction $h \in X$, then F is said to be *Gâteaux differentiable* at x.

Definition 1.18. Let X, Y be Banach spaces and let $\operatorname{dom}(F) \subset X$ be open. An operator $F : \operatorname{dom}(F) \to Y$ is called *Fréchet differentiable* at $x \in \operatorname{dom}(F)$ if there exists a bounded, linear operator $F'(x) : X \to Y$ such that

$$\lim_{h \to 0} \frac{\|F(x+h) - F(x) - F'(x)(h)\|_Y}{\|h\|_X} = 0.$$

Proposition 1.19. *If F is Fréchet differentiable at $x \in \text{dom}(F)$, then F is also Gâteaux differentiable at x and for every $h \in X$*

$$dF(x;h) = F'(x)(h)$$

holds.

If F is Fréchet differentiable at every point $x \in \text{dom}(F)$, then the Fréchet derivative is an operator

$$F' : \text{dom}(F) \to L(X,Y),$$

where $L(X,Y)$ denotes the Banach space of bounded, linear operators between X and Y equipped with the operator norm

$$\|T\|_{L(X,Y)} = \sup_{x \neq 0} \frac{\|Tx\|_Y}{\|x\|_X}.$$

If the operator F' is itself Fréchet differentiable at $x \in \text{dom}(F)$ then the second order Fréchet derivative of F at x is a bounded, linear operator

$$F''(x) : X \to L(X,Y)$$

and if the same holds true for every $x \in \text{dom}(F)$, then F'' is an operator

$$F'' : \text{dom}(F) \to L\big(X, L(X,Y)\big).$$

For simplicity and clarity of exposition the space $L\big(X, L(X,Y)\big)$ can be identified with the space of bilinear mappings $L^2(X \times X, Y)$, where an element $\mathcal{A} \in L\big(X, L(X,Y)\big)$ is identified with $A \in L^2(X \times X, Y)$ such that for all $g, h \in X$

$$\mathcal{A}(g)(h) = A(g,h).$$

Continuing this process we define higher order Fréchet derivatives recursively as follows.

Definition 1.20. Let X, Y be Banach spaces and let $\text{dom}(F) \subset X$ be open. An operator $F : \text{dom}(F) \to Y$ is called *n-times Fréchet differentiable* on $\text{dom}(F)$ if there exists an operator

$$F^{(n)} : \text{dom}(F) \to L^n(X^n, Y)$$

which is the Fréchet derivative of $F^{(n-1)}$. Here $L^n(X^n, Y)$ denotes the space of continuous, *multi-linear* mappings from X^n to Y.

If $\phi : \text{dom}(\phi) \subset X \to \mathbb{R}$ is a functional on X, then its Fréchet derivative at a point $x \in \text{dom}(F)$ is an element in the dual space $X^* = L(X, \mathbb{R})$. For convex functionals this notion of differentiability can be weakened to subdifferentials, which are well-defined even if the functional is discontinuous.

Definition 1.21. An element $\xi \in X^*$ is called a *subderivative* of the convex functional $\phi : \text{dom}(\phi) \subset X \to \mathbb{R}$ at a point $x \in \text{dom}(\phi)$, if for all $z \in \text{dom}(\phi)$ it holds that
$$\phi(z) - \phi(x) \geq \langle \xi, z - x \rangle.$$
The set of all subderivatives ξ at a point x is called the *subdifferential* and denoted by
$$\partial \phi(x) = \left\{ \xi \in X^* \mid \forall z \in \text{dom}(\phi) : \phi(z) - \phi(x) \geq \langle \xi, x - z \rangle \right\}.$$
We will make use of the following property of the subdifferential at a later point.

Proposition 1.22. *The subdifferential of a convex functional ϕ is a convex and closed set.*

For a convex functional every critical point is necessarily also a global minimizer. Therefore the necessary first order condition for a local minimum is also sufficient for a global minimum. This statement still holds true for subdifferentials.

Proposition 1.23. *A convex functional $\phi : \text{dom}(\phi) \subset X \to \mathbb{R}$ has a global minimum at $x \in \text{dom}(\phi)$ if and only if $0 \in \partial \phi(x)$.*

Throughout the later chapters we will work with minimizers of functionals that are sums of a differentiable and a convex term. For such functionals the following notion of a generalized derivative is well-defined.

Definition 1.24. Let $d : \text{dom}(d) \to \mathbb{R}$ be differentiable and $\phi : \text{dom}(\phi) \to \mathbb{R}$ be convex. Then the *generalized derivative* of
$$j : \text{dom}(d) \cap \text{dom}(\phi) \to \mathbb{R}, j(x) = d(x) + \phi(x)$$
at a point x is given by the set
$$\partial j(x) = d'(x) + \partial \phi(x).$$

We may again formulate the necessary first order condition, which is, however, no longer sufficient for a global minimizer.

Proposition 1.25. *Let $j(x) = d(x) + \phi(x)$ with $d(x)$ differentiable and $\phi(x)$ convex. If $j(x)$ has a global minimum at $x \in \text{dom}(d) \cap \text{dom}(\phi)$, then $0 \in \partial j(x)$.*

We conclude this section with the definitions of lower semi-continuity with respect to some topology and coercivity. These concepts are known to have a strong impact in variational calculus.

Definition 1.26. Let X be a Banach space. A functional

$$\phi : \text{dom}(\phi) \subset X \to \mathbb{R}$$

is called *weakly (weak-*) sequentially lower semi-continuous* if for any sequence $\{x_n\}$ in $\text{dom}(\phi)$ such that $x_n \rightharpoonup x \in X$ (resp. $x_n \rightharpoonup^* x \in X$) it holds that $x \in \text{dom}(\phi)$ and

$$\phi(x) \leq \liminf_{n \to \infty} \phi(x_n).$$

The following proposition is a consequence of the convexity and the continuity of the norm. Indeed, a convex and lower semi-continuous functional is also weakly lower semi-continous.

Proposition 1.27. *In any Banach space X the norm is weakly sequentially lower semi-continuous.*

Note that this is in general not true with respect to the weak-* topology.

Definition 1.28. Let X be a Banach space. A functional $\phi : X \to \mathbb{R}$ is called *weakly coercive* if for any unbounded sequence $\{x_n\} \in X^{\mathbb{N}}$, $\|x_n\| \to \infty$, it holds that

$$\phi(x_n) \to \infty \text{ as } n \to \infty.$$

We conclude with a simple, but useful properties of monomials on $\mathbb{R}^+$.

Lemma 1.29. *The mappings*

$$f_p : \mathbb{R}_0^+ \to \mathbb{R}_0^+, \quad f_p(s) = s^p$$

are convex for any $p \in [1, \infty)$.

Proof. Observe that the mappings $f_p(s)$ are twice differentiable on $\mathbb{R}^+$ with

$$f_p'(x) = ps^{p-1}, \qquad f_p''(x) = p(p-1)s^{p-2}.$$

Now, if $p = 1$, then $f_1(s)$ is the identity which is trivially convex. On the other hand, for $p > 1$ the second derivative is positive everywhere on $\mathbb{R}^+$, which shows that f_p is convex there. It remains to check that the convexity extends to $s = 0$ which is clearly the case. □

Chapter 2

Tikhonov regularization

We are interested in obtaining approximate solutions of an operator equation

$$F(x) = y, \qquad (2.1)$$

from *noisy data* y^δ satisfying $\|y - y^\delta\| \leq \delta$. We will refer to the bound $\delta > 0$ as the *noise level*. If problem (2.1) is ill-posed, then one needs to employ some sort of regularization to stabilize the process of (approximately) obtaining x from y^δ. Introducing the notation 2^X for the power set of X, i.e.,

$$2^X := \{A : A \subset X\},$$

the definition of a regularization method, conceptually, reads as follows (compare to [17, Definition 3.1]).

Definition 2.1. Let $F^\dagger : \mathrm{rg}(F) \subset Y \longrightarrow X$ be an operator that maps $y \in \mathrm{rg}(F)$ to some solution $x^\dagger = F^\dagger(y)$ of (2.1) (for example, one with minimal norm). A family of operators $R_\alpha : Y \to 2^X$ for $0 < \alpha \leq \alpha_0$, is called a *regularization* for $F^\dagger$ if

(i) *(Existence)* the set $R_\alpha(y^\delta)$ is non-empty for all $y^\delta \in Y$,

(ii) *(Uniqueness)* the set $R_\alpha(y^\delta)$ has at most one element for all $y^\delta \in Y$,

(iii) *(Stability)* the mappings R_α are continuous for every $\alpha > 0$,

(iv) *(Convergence)* for $y^\delta \in Y$ such that $\|y^\delta - y\| \leq \delta$ and suitably chosen $\alpha = \alpha(\delta, y^\delta)$ with

$$\alpha(\delta, y^\delta) \to 0$$

as $\delta \to 0$, it holds (in some topology) that

$$R_\alpha(y^\delta) \to F^\dagger(y)$$

as $\delta \to 0$.

Our notation $F^\dagger$ resembles the notation of the Moore-Penrose generalized inverse of a matrix. Indeed, in case of F being a bounded, linear operator the generalized inverse is not only an admissible choice for $F^\dagger$ in Definition 2.1, but historically was the first one to be studied in great detail. It can thus be considered a prototype for the more general operators $F^\dagger$ that are of interest to us here.

Note, that for linear operators $\mathrm{rg}(F)$ is a subspace of Y so that the domain of definition of $F^\dagger$ can be extended to $\mathrm{rg}(F) \oplus U$, where U is the largest subspace of Y such that $U \cap \mathrm{rg}(F) = \{0\}$. If Y is a Hilbert space, then $U = \mathrm{rg}(F)^\perp$ and

$$\mathrm{dom}(F^\dagger) = \mathrm{rg}(F) \oplus \mathrm{rg}(F)^\perp$$

is a dense subset of Y. For ill-posed operators $\mathrm{rg}(F)$ is not closed so that $\mathrm{dom}(F^\dagger)$ will not coincide with Y.

Remark 2.2. The Tikhonov method introduced in Definition 2.13 below chooses the regularized solutions $R_\alpha(y^\delta)$ in Definition 2.1 as minimizers of a variational functional. These minimizers will, however, not necessarily be unique, so that we will have to dispense with property (ii) in Definition 2.1. Note, that for set-valued functions R_α it is then not immediately clear how to define continuity in (iii) and convergence of $R_\alpha(y^\delta) \to F^\dagger(y)$ in (iv). Especially, since $F^\dagger(y)$ will itself be a set of solutions $x^\dagger$ of (2.1) that minimize a *penalty functional* $\Psi(x)$. Depending on Ψ, F and y there might thus be more than one $x^\dagger$ with the same value $\Psi(x^\dagger)$ such that $F(x^\dagger) = y$. In this case $F^\dagger$ is an operator $F^\dagger : \mathrm{rg}(F) \to 2^X$. ∎

To address this issue we thus need to clarify what we mean by convergence of a sequence $\{A_n\}$ of sets to a set A. This is done using cuts of the sets A_n in combination with the notion of convergence of a sequence to a set in Definition 1.12.

Definition 2.3. A sequence $\{a_n\}_{n \in \mathbb{N}}$ is called a *cut* of the family of sets $\{A_n\}_{n \in \mathbb{N}}$ if $a_n \in A_n$ holds for all n.

Then the sets A_n are said to converge to A with respect to a topology τ, if $a_n \to A$ with respect to τ in the sense of Definition 1.12 for every cut $\{a_n\}$ of $\{A_n\}$. We will see in Chapter 3 that the Tikhonov method is a regularization according to this notion of set-valued convergence.

2.1 Formulation of the Tikhonov method

The conditions in this section summarize our standing assumptions on the spaces X, Y, the operator F and the penalty term Ψ under consideration as well as the noisy data y^δ. These assumptions are in place throughout this work and go without saying.

Condition 2.4. Let X be the dual of some Banach space Z, $X = Z^*$, and let Y be a Banach space. Assume that

$$F : \text{dom}(F) \subset X \to Y$$

is an operator from a weak-* sequentially closed domain $\text{dom}(F)$ with $0 \in \text{dom}(F)$ into Y and that F is weak-* to weak continuous, i.e., if x_n is a sequence in $\text{dom}(F)$ and $x_n \rightharpoonup^* x \in X$, then $x \in \text{dom}(F)$ and $F(x_n) \rightharpoonup F(x)$ in Y.

Example 2.5. We give two examples of situations where Condition 2.4 is satisfied.

- If F is bounded, linear and

$$\text{rg}(F^*) \subset Z, \qquad (2.2)$$

where we identify Z with a subspace of its bidual X^* according to Proposition 1.5, then for every $\phi \in Y^*$ we have $F^*\phi \in Z$. Thus $\varphi_n \rightharpoonup^* \varphi$ implies that $F\varphi_n \rightharpoonup F\varphi$ since

$$\langle \phi, F\varphi_n \rangle = \langle F^*\phi, \varphi_n \rangle \to \langle F^*\phi, \varphi \rangle = \langle \phi, F\varphi \rangle.$$

holds for every $\phi \in Y^*$. Therefore any linear operator satisfying (2.2) is weak-* to weak continuous.

- If X is reflexive, then $Z = X^*$ and for any linear F (2.2) clearly holds. Since according to Proposition 1.10 the weak and weak-* topology coincide on reflexive Banach spaces, we require that the operator is weak-weak continuous. The autoconvolution operator which we study in Chapter 8 is an example of a non-linear operator which fulfills this assumption.

Lemma 2.6. *If Condition 2.4 is satisfied, then for any $y \in Y$ the functional on X*

$$G : x \mapsto \|F(x) - y\|$$

is weak- sequentially lower semicontinuous.*

Proof. Let $x_n \rightharpoonup^* x$ in X, then according to Condition 2.4 it also holds that $F(x_n) \rightharpoonup F(x)$ in Y and consequently $F(x_n) - y \rightharpoonup F(x) - y$. The result thus follows readily from the weak sequential l.s.c. of the norm (cf. Proposition 1.27). □

Having established the topological setup of the problem under consideration, we now define the solution operator $F^\dagger$ from Definition 2.1 that we would like to approximate. To this end we restrict our attention to solutions $x^\dagger$ of (2.1) that share specific properties, namely, that they minimize a so-called *penalty functional* $\Psi(x)$ (compare to Remark 2.2).

Condition 2.7. Let $\Psi : X \to \mathbb{R}_0^+ \cup \{+\infty\}$ be a convex functional with
$$\mathrm{dom}(\Psi) = \{x \in X \mid \Psi(x) < \infty\}$$
and $0 \in \mathrm{dom}(\Psi)$ such that

(i) $\Psi(x) = 0$ if and only if $x = 0$,

(ii) Ψ is weak-* sequentially lower semi-continous,

(iii) Ψ is weakly coercive.

Note that if $\tilde{\Psi}(x)$ fulfills Condition 2.7 except that it takes the value zero only in $x^* \neq 0$, then we may use the shifted functional $\Psi(x) = \tilde{\Psi}(x - x^*)$ instead.

An immediate consequence of the properties of Ψ is the following lemma.

Lemma 2.8. *If Ψ satisfies Condition 2.7, then for any sequence $\{x_n\} \subset X$ with $\Psi(x_n) \to 0$ it holds that $x_n \rightharpoonup^* 0$.*

Proof. Take an arbitrary subsequence of $\{x_n\}$ – again denoted by $\{x_n\}$ for simplicity – then $\{\Psi(x_n)\}$ is clearly bounded since it goes to zero and because of the weak coercivity of Ψ, so is $\{x_n\}$. Therefore we can extract a weak-* convergent subsequence, $x_{n'} \rightharpoonup^* \bar{x}$. Due to the weak-* lower semi-continuity of Ψ we obtain
$$0 \leq \Psi(\bar{x}) \leq \liminf_{n' \to \infty} \Psi(x_{n'}) = 0,$$
which according to Condition 2.7 (i) only holds for $\bar{x} = 0$.

Altogether we have shown that any subsequence of $\{x_n\}$ has a subsequence that converges weak-* to 0 and therefore the same holds true for the entire sequence. □

In the preceding proof we have used a well known convergence principle in Banach spaces, which is explained in detail in [66, Proposition 10.13]. We now state two slightly different versions of this convergence principle, which we will need later on.

Lemma 2.9. *Let $\{x_n\} \subset X$ and a functional $f : X \to \mathbb{R}$ be such that every subsequence $\{x_{n'}\}$ of $\{x_n\}$ has, in turn, a subsequence $\{x_{n''}\}$ such that $f(x_{n''}) \to c \in \mathbb{R}$ as $n'' \to \infty$, then $f(x_n) \to c$ as $n \to \infty$.*

Proof. If $f(x_n) \to c$ does not hold then there is a subsequence $\{x_{n'}\}$ such that $|f(x_{n'}) - c| > \varepsilon$ for some $\varepsilon > 0$. This contradicts the assumption that $\{x_{n'}\}$ has a subsequence $\{x_{n''}\}$ such that $f(x_{n''}) \to c$. □

The proof of the second version is based on the same idea only applied to a different topological set-up.

Lemma 2.10. *Let $\{x_n\} \subset X$ and $A \subset X$ be such that every subsequence $\{x_{n'}\}$ of $\{x_n\}$ has, in turn, a subsequence $\{x_{n''}\}$ such that $x_{n''} \rightharpoonup^* x \in A$ as $n'' \to \infty$, then $x_n \rightharpoonup^* A$ as $n \to \infty$ in the sense of Definition 1.12.*

Proof. If $x_n \rightharpoonup^* A$ does not hold then there exists a neighbourhood T of A in the weak-* topology and a subsequence $\{x_{n'}\}$ such that $x_{n'} \notin T$ for all $n' \in \mathbb{N}$. This contradicts the assumption that $\{x_{n'}\}$ has a subsequence $\{x_{n''}\}$ which converges weak-* to some element of A. □

We now define the notion of a Ψ-minimizing solution.

Definition 2.11. *A solution $x^\dagger$ of $F(x) = y$ is called a Ψ-minimizing solution if*
$$\Psi(x^\dagger) = \min\{\Psi(x) : x \in X, F(x) = y\} < \infty.$$

We define the operator $F^\dagger$ via
$$F^\dagger : \mathrm{rg}(F) \subset Y \to 2^X, \ y \mapsto F^\dagger(y) = \mathcal{L},$$

where $\mathcal{L}$ denotes the set of all Ψ-minimizing solutions, i.e.

$$\mathcal{L} = \left\{ x^\dagger \in X \mid F(x^\dagger) = y \text{ and } \Psi(x^\dagger) \leq \Psi(x) \ \forall x : F(x) = y \right\}. \quad (2.3)$$

Throughout this work we assume $\mathcal{L} \neq \emptyset$ and write $\psi^\dagger$ for the common value of the penalty functional Ψ evaluated at any $x^\dagger \in \mathcal{L}$, i.e.,

$$\psi^\dagger = \Psi(x^\dagger) < \infty, \qquad x^\dagger \in \mathcal{L}. \quad (2.4)$$

The following Lemma sheds some light on the assumption $L \neq \emptyset$ in Definition 2.11.

Lemma 2.12. *If there exists a solution $\bar{x}$ of $F(x) = y$ that belongs to $\mathrm{dom}(\Psi)$, then there also exists a Ψ-minimizing solution $x^\dagger$ of $F(x) = y$.*

Proof. Let
$$\psi^\dagger = \inf\left\{ \Psi(x) \mid x \in \mathrm{dom}(F), \ F(x) = y \right\},$$
then
$$0 \leq \psi^\dagger \leq \Psi(\bar{x}) < \infty$$

as $\bar{x} \in \mathrm{dom}(\Psi)$ and $\Psi(x) \geq 0$ for all $x \in X$. Thus, there exists a minimizing sequence $\{x_n\} \subset \mathrm{dom}(F)$ such that $F(x_n) = y$ for all $n \in \mathbb{N}$ and

$$\Psi(x_n) \to \psi^\dagger.$$

Without loss of generality we furthermore assume that $\Psi(x_n) \leq \Psi(\bar{x}) < \infty$ for all $n \in \mathbb{N}$. Due to the weak coercivity assumption on Ψ in Condition 2.7

it follows that $\{x_n\}$ is bounded in X. Since X is the dual of some Banach space Z (cf. Condition 2.4) Proposition 1.8 asserts that $\{x_n\}$ contains a weak-* convergent subsequence $\{x_k\} \subset \{x_n\}$, $x_k \rightharpoonup^* x^\dagger \in X$. It remains to show that $x^\dagger$ is indeed a Ψ-minimizing solution. The weak-* l.s.c. of $G(x)$ in Lemma 2.6 and of $\Psi(x)$ in Condition 2.7, respectively, yield

$$\left\| F(x^\dagger) - y \right\| \leq \liminf_{k \to \infty} \left\| F(x_k) - y \right\| = 0$$

and

$$\Psi(x^\dagger) \leq \liminf_{k \to \infty} \Psi(x_k) = \psi^\dagger.$$

Therefore, $x^\dagger \in \mathcal{L}$ as defined in (2.3). □

At this point we have explained the solutions $x^\dagger \in \mathcal{L}$ that we are looking for in terms of the penalty functional $\Psi(x)$. The method that we will use to stably approximate these solutions is known as *Tikhonov regularization*. It is defined as follows.

Definition 2.13. Let F, Ψ fulfill Conditions 2.4 and 2.7, respectively. For $\alpha > 0$ and $y^\delta \in Y$ the *variational Tikhonov functional* is defined as

$$J_{\alpha, y^\delta}(x) = \begin{cases} \left\| F(x) - y^\delta \right\|^q + \alpha \Psi(x) & \text{if } x \in \mathcal{D} \\ +\infty & \text{otherwise,} \end{cases} \qquad (2.5)$$

where $q > 0$ is fixed and

$$\mathcal{D} := \mathrm{dom}(F) \cap \mathrm{dom}(\Psi). \qquad (2.6)$$

The family of *Tikhonov regularization* operators $\{R_\alpha\}$ is given by

$$R_\alpha : Y \to 2^X, \ y^\delta \mapsto R_\alpha(y^\delta) = \mathcal{M}_{\alpha, y^\delta},$$

where

$$\mathcal{M}_{\alpha, y^\delta} = \arg\min_{x \in X} \left\{ J_{\alpha, y^\delta}(x) \right\}. \qquad (2.7)$$

Hence, the *regularized solutions* $x_\alpha^\delta \in \mathcal{M}_{\alpha, y^\delta}$ are the minimizers of the functionals $J_{\alpha, y^\delta}(x)$.

We will see in Chapter 3 that the family of operators $\{R_\alpha\}_{\alpha > 0}$ is a convergent regularization method for $\mathcal{L} = F^\dagger(y)$ as defined in (2.3) in the sense of Definition 2.1. However, as mentioned in Remark 2.2, in general, we have to dispense with the uniqueness of $x_\alpha^\delta \in R_\alpha(y^\delta)$ and use a set-valued definition of convergence and continuity. But before getting to this, we mention a few properties of Tikhonov regularization and discuss several examples of penalty terms that fit within the framework described above.

Remark 2.14. When using Tikhonov regularization it would suffice to assume the sequential continuity of F in Condition 2.4 on sublevelsets $S_\alpha(\sigma)$ of $J_{\alpha,y}(x)$,
$$S_\alpha(\sigma) = \{x \in X \mid J_{\alpha,y}(x) \leq \sigma\},$$
for $\alpha > 0$ and $\sigma > 0$. This, in turn, is implied by the weak-* to weak sequential closedness of F according to Definition 1.15 provided that Y is reflexive: Adapting the arguments in [56, p. 61] to our situation, we take a weak-* convergent sequence $\{x_n\}_{n\in\mathbb{N}} \subset S_\alpha(\sigma)$ with limit $x \in X$ and find
$$\|F(x_n) - y\|^q \leq J_{\alpha,y}(x_n) \leq \sigma$$
for all $n \in \mathbb{N}$. Therefore the sequence $\{F(x_n)\}_{n\in\mathbb{N}}$ is bounded in the reflexive space Y and we may extract a weakly convergent subsequence $\{F(x_k)\}$. From the weak-* to weak sequential closedness of F we obtain that $x \in \text{dom}(F)$ and $F(x_k) \rightharpoonup F(x)$. Due to the weak-* sequential l.s.c. of $G(x)$ in Lemma 2.6 and $\Psi(x)$ in Condition 2.7
$$J_{\alpha,y}(x) \leq \liminf_{n\to\infty} J_{\alpha,y}(x_n) \leq \sigma$$
so that $x \in S_\alpha(\sigma)$. The weak-* to weak continuity of F now follows using Lemma 2.9 as these arguments apply to any subsequence of the original sequence $\{x_n\}$. ∎

If Y is a Hilbert space and the operator F is Fréchet differentiable, then we can formulate the first order necessary condition to characterize the minimizers x_α^δ of (2.5) by means of the generalized derivative in Definition 1.24.

Lemma 2.15. *Assume that Y is a Hilbert space and that the operator F is Fréchet differentiable. Then the first order necessary condition in X^* for a minimizer x_α^δ of the Tikhonov functional (2.5) reads*
$$q \left\|F(x_\alpha^\delta) - y^\delta\right\|^{q-2} F'(x_\alpha^\delta)^* \left(y^\delta - F(x_\alpha^\delta)\right) \in \alpha \partial \Psi(x_\alpha^\delta). \tag{2.8}$$

In general, the minimizers of (2.5) will not be unique. Strictly speaking, the neighbouring problems
$$\min_{x \in X} J_{\alpha,y^\delta}(x)$$
therefore remain ill-posed in the sense of Hadamard because the solution may not be unique. We will show, however, that any choice of a minimizer $x_\alpha^\delta \in \mathcal{M}_{\alpha,y^\delta}$ provides a regularization method that converges to the set $\mathcal{L}$ of Ψ-minimizing solutions. For our purposes we assume that all Ψ-minimizing solutions of (2.1) are equally acceptable to us.

We will frequently encounter the following examples of penalty terms, which fulfill the Condition 2.7.

2.2 ℓ_p-norms of frame coefficients in Hilbert spaces

The last decade has seen a growing interest in sparsity promoting regularization, which serves as one of the main motivations for our analysis. The idea is to choose the penalty term as a (weighted) ℓ_p-norm with $p < 2$ of the coefficients with respect to some frame in X, which is defined as follows.

Definition 2.16. Let X be a Hilbert space. A collection $\Phi = \{\phi_\lambda\}_{\lambda \in \Lambda}$ with a index set Λ is called a *frame* for X, if there exist constants $A, B > 0$ such that
$$A \|x\|^2 \leq \sum_{\lambda \in \Lambda} |\langle \phi_\lambda, x \rangle|^2 \leq B \|x\|^2.$$

If the frame bounds A, B satisfy $A = B$, then the frame is said to be *tight*.

Straight forward examples of frames are (orthonormal) bases of Hilbert spaces or unions thereof. Using collections of elements ϕ_λ that are not linearly independent and therefore include some redundancy facilitates the existence of points $x \in X$ that are represented as a linear combination of only few ϕ_λ. However, there may no longer be a 1-to-1 correspondence between points in X and their coefficient sequences with respect to a frame Φ.

Let us now define the sparsity promoting weighted ℓ_p-norms of the frame coefficients.

Definition 2.17. Let $\Phi = \{\phi_\lambda\}_{\lambda \in \Lambda}$ be a frame for the Hilbert space X. For a fixed sequence $w = \{w_\lambda\}_{\lambda \in \Lambda}$ which is assumed to be bounded from below away from zero, $0 < w_0 \leq w_\lambda$ for all $\lambda \in \Lambda$ and for $1 \leq p \leq 2$, we define

$$\Psi_{p,w}(x) = \sum_{\lambda \in \Lambda} w_\lambda |\langle \phi_\lambda, x \rangle|^p, \qquad (2.9)$$

with
$$\text{dom}(\Psi_{p,w}) = \{x \in X : \Psi_{p,w}(x) < \infty\}.$$

In order to use the functionals $\Psi_{p,w}$ as penalty terms in the Tikhonov functional (2.5) and apply our theory we check the requirements in Condition 2.7.

Lemma 2.18. *The functionals $\Psi_{p,w}$ defined in (2.9) are convex on their domains of definition for $1 \leq p \leq 2$.*

Proof. Let $t \in (0, 1)$ and $x, z \in \text{dom}(\Psi_{p,w})$, then using the convexity of

$f_p(s)$ from Lemma 1.29 we find

$$\Psi_{p,w}(tx + (1-t)z) = \sum_{\lambda \in \Lambda} w_\lambda \, |\langle \phi_\lambda, tx + (1-t)z \rangle|^p$$
$$\leq \sum_{\lambda \in \Lambda} w_\lambda \, (t \, |\langle \phi_\lambda, x \rangle| + (1-t) \, |\langle \phi_\lambda, z \rangle|)^p$$
$$\leq t \sum_{\lambda \in \Lambda} w_\lambda \, |\langle \phi_\lambda, x \rangle|^p + (1-t) \sum_{\lambda \in \Lambda} w_\lambda \, |\langle \phi_\lambda, z \rangle|^p$$
$$= t\Psi_{p,w}(x) + (1-t)\Psi_{p,w}(z).$$

$\square$

To show that the functionals $\Psi_{p,w}$ fulfill the assumptions in Condition 2.7 we recall another basic property of the ℓ_p spaces.

Lemma 2.19. *Let $0 < p \leq q < \infty$, then for any sequence $x = \{x_\lambda\}_{\lambda \in \Lambda} \in \ell_p(\Lambda)$ it holds*
$$\|x\|_{\ell_q(\Lambda)} \leq \|x\|_{\ell_p(\Lambda)},$$
which especially shows that $\ell_p(\Lambda) \subset \ell_q(\Lambda)$.

Proof. Clearly, for any $\lambda \in \Lambda$,
$$\frac{|x_\lambda|}{\|x\|_{\ell_p(\Lambda)}} \leq 1$$
and we thus see that
$$\left(\frac{|x_\lambda|}{\|x\|_{\ell_p(\Lambda)}} \right)^q \leq \left(\frac{|x_\lambda|}{\|x\|_{\ell_p(\Lambda)}} \right)^p \leq 1.$$
Summing over all $\lambda \in \Lambda$ we get
$$\left(\frac{\|x\|_{\ell_q(\Lambda)}}{\|x\|_{\ell_p(\Lambda)}} \right)^q = \sum_{\lambda \in \Lambda} \frac{|x_\lambda|^q}{\|x\|_{\ell_p(\Lambda)}^q} \leq \sum_{\lambda \in \Lambda} \frac{|x_\lambda|^p}{\|x\|_{\ell_p(\Lambda)}^p} = 1,$$
which yields
$$\|x\|_{\ell_q(\Lambda)} \leq \|x\|_{\ell_p(\Lambda)}.$$

$\square$

From Lemma 2.19 we obtain the strong coercivity of $\Psi_{p,w}(x)$.

Lemma 2.20. *If $\Psi_{p,w}(x)$ is as in Definition 2.17 with given $\Phi = \{\phi_\lambda\}_{\lambda \in \Lambda}$, $w = \{w_\lambda\}_{\lambda \in \Lambda}, w_0$ and p, then*

$$\|x\|^p \leq \frac{1}{w_0 \sqrt{A^p}} \Psi_{p,w}(x) \qquad (2.10)$$

holds for all $x \in X$, where A denotes the lower frame bound of Φ.

Proof. From Definition 2.16 of the frame Φ and Lemma 2.19 with $q = 2$ we obtain

$$\|x\| \leq \frac{1}{\sqrt{A}} \left(\sum_{\lambda \in \Lambda} |\langle \phi_\lambda, x \rangle|^2 \right)^{1/2}$$

$$\leq \frac{1}{\sqrt{A}} \left(\sum_{\lambda \in \Lambda} |\langle \phi_\lambda, x \rangle|^p \right)^{1/p}$$

$$\leq \frac{1}{w_0^{1/p} \sqrt{A}} \left(\sum_{\lambda \in \Lambda} w_\lambda |\langle \phi_\lambda, x \rangle|^p \right)^{1/p},$$

which gives the assertion when taken to the p-th power. $\square$

With these results we are in position to show that weighted ℓ_p penalization is admissible according to Condition 2.7.

Lemma 2.21. *Let $\Phi = \{\phi_\lambda\}_{\lambda \in \Lambda}$ be a frame for the Hilbert space X and let $w = \{w_\lambda\}_{\lambda \in \Lambda}$ be such that $0 < w_0 \leq w_\lambda$ for all $\lambda \in \Lambda$. For $1 \leq p \leq 2$ the functional $\Psi_{p,w}(x)$ defined as in (2.9) satisfies Condition 2.7.*

Proof. Noting that $\langle \phi_\lambda, 0 \rangle = 0$ for all $\lambda \in \Lambda$ we obtain that $\Psi_{p,w}(0) = 0$ which entails that $0 \in \text{dom}(\Psi_{p,w})$. For $x \neq 0$ we see from (2.10) that $\Psi_{p,w}(x) > 0$, which shows that $\Psi_{p,w}(x) = 0$ if and only if $x = 0$.

Also the weak coercivity is an immediate consequence of (2.10) as

$$w_0 \sqrt{A^p} \|x\|^p \to \infty \quad \text{as} \quad \|x\| \to \infty$$

and therefore so does $\Psi_{p,w}(x)$.

For the weak-* lower semicontinuity we first note that every Hilbert space is reflexive and that thus – according to Lemma 1.10 – the weak and weak-* topology on X coincide. Now, to show weak l.s.c. we take the proof from [56, p. 80], where it is argued that the weak continuity of the functionals $x \mapsto |\langle x, \phi_\lambda \rangle|^p$ implies that for every weakly convergent sequence $x_n \rightharpoonup x$ and all $\lambda \in \Lambda$

$$\lim_{n \to \infty} w_\lambda |\langle x_n, \phi_\lambda \rangle|^p = w_\lambda |\langle x, \phi_\lambda \rangle|^p.$$

It then follows from Fatou's Lemma (see, e.g., [56, Lemma 9.8]) that

$$\Psi_{p,w}(x) = \sum_{\lambda \in \Lambda} w_\lambda |\langle x, \phi_\lambda \rangle|^p$$

$$= \sum_{\lambda \in \Lambda} \liminf_{n \to \infty} w_\lambda |\langle x_n, \phi_\lambda \rangle|^p$$

$$\leq \liminf_{n \to \infty} \sum_{\lambda \in \Lambda} w_\lambda |\langle x_n, \phi_\lambda \rangle|^p$$

$$= \liminf_{n \to \infty} \Psi_{p,w}(x_n).$$

We will study the sparsity promoting regularization with the weighted ℓ_p-norms $\Psi_{p,w}(x)$ again in Chapter 7.

2.3 Classical Tikhonov regularization in Banach spaces

Variational regularization methods such as the formulation (2.5) considered here were first studied in a Hilbert space setting with penalty term $\Psi(x) = \|x\|^2$. The requirements on Ψ in Condition 2.7 are, however, also met in reflexive Banach spaces.

Proposition 2.22. *If X is a reflexive Banach space and $p \geq 1$, then the functional $\Psi(x) = \|x\|_X^p$ with $\mathrm{dom}(\Psi) = X$ fulfills Condition 2.7.*

Proof. Using the norm properties we see that $\Psi(x) = 0$ if and only if $x = 0$.

According to Proposition 1.10 the weak and weak-* topologies agree on reflexive Banach spaces and thus the norm on X is weak-* sequentially l.s.c. due to Proposition 1.27. But then for $x_n \rightharpoonup^* \bar{x} \in X$, clearly,

$$\Psi(x) = \|x\|^p \leq \liminf_{n \to \infty} \|x_n\|^p = \Psi(x_n).$$

The weak coercivity is just as easily obtained for this choice of Ψ, since $\Psi(x) = \|x\|^p \to \infty$ as $\|x\| \to \infty$. □

2.4 Besov spaces

Our first example of a Banach space regularized problem that meets the imposed conditions is formulated in *Besov spaces*. They are subspaces of the space of tempered distributions $\mathcal{S}'(\mathbb{R}^d)$ that are commonly defined via the modulus of smoothness. We refer the interested reader to [16] for a detailed introduction to Besov spaces and to [43] for regularization results with a-priori parameter choice rules in Besov spaces.

For our purposes it suffices to know that for $s \in \mathbb{R}$ and $p \geq 1$ the Besov spaces $B_p^s(\mathbb{R}^d)$ are Banach spaces (cf. [53]), which are reflexive for $p > 1$, and that according to [11] they can be characterized by means of wavelets. We restrict ourselves to a rudimentary definition of wavelets here and refer the interested reader to [12] for a detailed introduction to the topic.

Definition 2.23. Let $\varphi_1, \ldots, \varphi_{2^d-1}$ be a sequence of *mother wavelets* such that the collection

$$\left\{ \varphi_{(i,j,k)} \mid 1 \leq i \leq 2^d - 1,\, j \in \mathbb{Z},\, k \in \mathbb{Z}^d \right\}$$

is an orthonormal basis of $L^2(\mathbb{R}^d)$, where

$$\varphi_{(i,j,k)}(x) = 2^{j/2}\varphi_i(2^j x - k), \quad x \in \mathbb{R}^d.$$

Assume that there exists an associated *scaling function* $\psi(x)$ such that

$$\overline{\bigoplus_{j \geq 0} W_j \oplus V} = L^2,$$

where

$$W_j = \text{span}\left\{\varphi_{(i,j,k)} \mid 1 \leq i \leq 2^d - 1, k \in \mathbb{Z}^d\right\},$$
$$V = \text{span}\left\{\psi(\,.\,-k) \mid k \in \mathbb{Z}^d\right\}.$$

It is a convenient slight abuse of notation to collect the wavelets and scaling functions in one *wavelet basis* $\{\phi_\lambda\}_{\lambda \in \Lambda}$, where the index set Λ consists of triples $\lambda = (i,j,k)$ with $1 \leq i \leq 2^d - 1$, $j \in \mathbb{Z}$, $k \in \mathbb{Z}^d$ for the wavelets and $\lambda = (0,0,k)$ with $k \in \mathbb{Z}^d$ which refer to the shifted scaling functions. We define $|\lambda|$ to be the scale j of the wavelet, which is 0 for the scaling functions.

Then, to every s, p there exists a wavelet basis $\{\phi_\lambda\}_{\lambda \in \Lambda}$ such that

$$\|x\|^p_{B^s_p} \sim \sum_{\lambda \in \Lambda} w_\lambda |x_\lambda|^p, \tag{2.11}$$

where

$$x_\lambda = \langle x, \phi_\lambda \rangle = \int_{\mathbb{R}^d} x(t)\phi_\lambda(t)dt \tag{2.12}$$

and the weights $w = \{w_\lambda\}_{\lambda \in \Lambda}$ are given by

$$w_\lambda = 2^{p\left(s + d\left(\frac{1}{2} - \frac{1}{p}\right)\right)|\lambda|}.$$

Note that for functionals $f, g : X \to \mathbb{R}$ we write $f \sim g$ if and only if there exist constants $0 < c \leq C$ such that

$$cg(x) \leq f(x) \leq Cg(x), \quad \forall x \in X.$$

Besov spaces coincide with Sobolev spaces for certain values of s and p. For example, it holds that $B^s_p = W^s_p$ for $s \notin \mathbb{Z}$ and $p \geq 1$ and that $B^s_2 = H^s = W^s_2$ for all $s \in \mathbb{R}$.

In the same way as in Section 2.3 we now obtain that Tikhonov regularization with a Besov space penalty term or the equivalent weighted ℓ_p-norm of the wavelet coefficients is admissible according to Condition 2.7.

Lemma 2.24. *Let $s \in \mathbb{R}$ and $p > 1$ and let $\{\phi_\lambda\}_{\lambda \in \Lambda}$ be a wavelet basis such that (2.11) holds. Then the choices $\Psi(x) = \|x\|_{B_p^s}^p$ and $\Psi(x) = \sum w_\lambda |x_\lambda|^p$ fulfill Condition 2.7.*

Proof. The assertion for $\Psi(x) = \|x\|_{B_p^s}^p$ is exactly Proposition 2.22 since B_p^s is reflexive for $p > 1$. For

$$\Psi(x) = \sum_{\lambda \in \Lambda} w_\lambda |x_\lambda|^p$$

with x_λ as defined in (2.12) we obtain that $\Psi(x) = 0$ if and only if $x = 0$ and the weak coercivity immediately from (2.11). The weak-* sequential l.s.c. follows as in Lemma 2.21 from the weak continuity of the functionals $x \mapsto |\langle x, \phi_\lambda \rangle|^p$ and Fatou's Lemma. □

At this point the theoretical requirements for Tikhonov regularization as defined in Section 2.1 are already met if the operator F maps $X = B_p^s$ to Y and fulfills Condition 2.4 provided that Ψ is chosen as in Lemma 2.24 for the same space $X = B_p^s$. However, due to the following continuous embedding result from [53], it is possible to also consider cases where the spaces on which F and Ψ are defined differ from each other.

Lemma 2.25. *Let $s_1, s_2 \in \mathbb{R}$ and $p_1 \leq p_2$. If*

$$s_1 - \frac{d}{p_1} > s_2 - \frac{d}{p_2},$$

then $B_{p_1}^{s_1} \hookrightarrow B_{p_2}^{s_2}$ continuously.

Therefore, an operator equation between a Besov space $X = B_{p_2}^{s_2}$ and a space Y can be regularized by a variational functional (2.5) with penalty terms $\Psi(x) = \|x\|_{B_{p_1}^{s_1}}^{p_1}$ on $\mathrm{dom}(\Psi) = B_{p_1}^{s_1}$ or a corresponding wavelet coefficient expansion $\Psi(x) = \sum w_\lambda |x_\lambda|^{p_1}$ on $\mathrm{dom}(\Psi) = B_{p_1}^{s_1}$ provided that s_1, s_2 and p_1, p_2 are as in Lemma 2.25.

2.5 TV regularization

As another important example which fulfills our theoretical assumptions we present total variation regularization, which has proven very successful for various tasks in mathematical imaging such as denoising, deblurring, segmentation, and inpainting. We refer the interested reader to [10] and the references therein for further details in this respect and also for the proofs of the results in this subsection.

Definition 2.26. Let $\Omega \subset \mathbb{R}^d$. For $x \in L^1(\Omega)$, we define the *total variation* of x as
$$TV(x) = \sup_{\substack{g \in C_0^\infty(\Omega; \mathbb{R}^d) \\ \|g\|_\infty \leq 1}} \int_\Omega x \operatorname{div}(g) \, dt,$$
where
$$\|g\|_\infty = \operatorname*{ess\,sup}_{t \in \Omega} \sqrt{g_1(t)^2 + \ldots + g_d(t)^2}, \qquad (2.13)$$
which gives the isotropic total variation. The space of bounded variation is then defined as
$$BV(\Omega) = \left\{ x \in L^1(\Omega) \mid TV(x) < \infty \right\}$$
with norm
$$\|x\|_{BV} = \|x\|_{L^1(\Omega)} + TV(x).$$

Other (anisotropic) functionals have been considered instead of (2.13), which lead to different reconstructions. If the function x is differentiable, then the total variation takes the following simpler form.

Proposition 2.27. *If $x \in W^{1,1}(\Omega)$, then*
$$TV(x) = \int_\Omega |\nabla x(t)| \, dt.$$

Our topological assumptions are satisfied on the subspace of BV with mean zero,
$$BV_0(\Omega) = \left\{ x \in BV(\Omega) \mid \int_\Omega x \, dt = 0 \right\}. \qquad (2.14)$$

For certain linear operators minimization over BV is indeed equivalent to minimization over BV_0. We will now identify the weak-* topology on BV_0.

Proposition 2.28. *Let*
$$Z_0 = \left\{ \operatorname{div}(g) \mid g \in C_0^\infty(\Omega; \mathbb{R}^d) \right\}$$
with norm
$$\|p\|_Z = \inf_{\substack{g \in C_0^\infty(\Omega; \mathbb{R}^d) \\ \operatorname{div}(g) = p}} \|g\|_{L^\infty}$$
and denote the completion of Z_0 in norm by
$$Z = \overline{Z_0},$$
then $BV_0(\Omega)$ can be identified with the dual space of Z.

From the definition of the total variation it is clear that $TV(0) = 0$.

Proposition 2.29. *The total variation is an equivalent norm on $BV_0(\Omega)$, i.e., there exist $c, C \in \mathbb{R}^+$ such that for all $x \in BV_0(\Omega)$*

$$c \left\| x \right\|_{BV} \leq TV(x) \leq C \left\| x \right\|_{BV}.$$

As an immediate consequence of Proposition 2.29 we obtain that in $BV_0(\Omega)$ there is no element other than zero for which the total variation vanishes and, moreover, the weak coercivity of TV.

Proposition 2.30. *The total variation is weak-* sequentially lower semicontinous on $BV_0(\Omega)$.*

Therefore all the properties in Condition 2.4 and 2.7 are satisfied for the total variation functional on the space $BV_0(\Omega)$ whenever F is any weak-* to weak continous operator from $BV_0(\Omega)$ into some Banach space Y.

Chapter 3
Regularization properties

In this chapter we study the regularization properties of approximating the Ψ-minimizing solutions $x^\dagger \in \mathcal{L}$ by minimizers $x_\alpha^\delta \in \mathcal{M}_{\alpha,y^\delta}$ of the Tikhonov-type functional (2.5). As introduced in Definitions 2.11 and 2.13 we thus regularize the operator
$$F^\dagger(y) = \mathcal{L}$$
using the family $\{R_\alpha\}_{\alpha>0}$ given by
$$R_\alpha(y^\delta) = \mathcal{M}_{\alpha,y^\delta},$$
where $\mathcal{L}$ and $\mathcal{M}_{\alpha,y^\delta}$ are as defined in (2.3) and (2.7), respectively. In other words, we replace the ill-posed problem of determining a solution x of $F(x) = y$ from noisy data y^δ by a sequence of neighbouring problems, namely, finding a minimizer of
$$J_{\alpha,y^\delta}(x) = \left\| F(x) - y^\delta \right\|^q + \alpha \Psi(x), \quad x \in \mathcal{D}.$$

In general, these regularized problems are, however, still not well-posed. Due to the non-linearity of the operator F under consideration, the regularized solution, i.e., the minimizer of the Tikhonov functional corresponding to some $\alpha > 0$ and $y^\delta \in Y$, might not be unique. Therefore, as mentioned in Remark 2.2, we will use a set-valued regularization approach and prove stability and convergence using cuts of a sequence of sets.

3.1 Existence and Stability

We will now show that the properties of the penalty term in Condition 2.7 ensure that a minimizer of the Tikhonov functional exists. This corresponds to the existence of a regularized solution in $R_\alpha(y^\delta) \neq \emptyset$ in Definition 2.1 (i). Such results have also been derived in [56, Section 3.2], for example.

Theorem 3.1. *If the operator F and the penalty term Ψ fulfill Conditions 2.4 and 2.7, then for any $y^\delta \in Y$ the set $\mathcal{M}_{\alpha,y^\delta}$ of minimizers of the Tikhonov functional $J_{\alpha,y^\delta}(x)$ defined in (2.5) is non-empty and $\mathcal{M}_{\alpha,y^\delta} \subset \mathcal{D}$.*

Proof. We know that $0 \in \mathcal{D} = \text{dom}(F) \cap \text{dom}(\Psi)$ according to Conditions 2.4 and 2.7. Therefore

$$0 \leq \inf_{x \in X} J_{\alpha,y^\delta}(x) \leq J_{\alpha,y^\delta}(0) < \infty$$

and in case minimizers exist, they thus necessarily belong to $\mathcal{D}$. Moreover, we may choose a sequence $\{x_n\}$ in $\mathcal{D}$ such that

$$\lim_{n \to \infty} J_{\alpha,y^\delta}(x_n) = \inf_{x \in X} J_{\alpha,y^\delta}(x).$$

It follows that $\{\Psi(x_n)\}$ is bounded and due to the weak coercivity of Ψ so is $\{x_n\}$. We may then extract a weak-* convergent subsequence $x_{n'} \rightharpoonup^* \bar{x}$. Using the weak-* l.s.c. of G in Lemma 2.6 and Ψ in Condition 2.7 we get

$$\begin{aligned} J_{\alpha,y^\delta}(\bar{x}) &= \left\| F(\bar{x}) - y^\delta \right\|^q + \alpha \Psi(\bar{x}) \\ &\leq \liminf_{n' \to \infty} \left\{ \left\| F(x_{n'}) - y^\delta \right\|^q + \alpha \Psi(x_{n'}) \right\} \\ &= \liminf_{n' \to \infty} J_{\alpha,y^\delta}(x_{n'}) = \inf_{x \in X} J_{\alpha,y^\delta}(x) \end{aligned}$$

so that we have found a minimizer $\bar{x} \in \mathcal{M}_{\alpha,y^\delta}$. □

The aspect which is most important in order to computationally obtain good approximations of a searched-for solution is stability. To formulate the stability result for the Tikhonov regularized family of operators R_α precisely, we consider an arbitrary cut of $\{R_\alpha(y_n)\}_{n \in \mathbb{N}}$ (cf. Definition 2.3) and use the notion of convergence of a sequence to a set that is given in Definition 1.12.

Theorem 3.2. *Let $y^\delta \in Y$, $y_n \to y^\delta$ and $\alpha > 0$ be fixed. If the operator F and the penalty term Ψ fulfill Conditions 2.4 and 2.7, then any choice of*

$$x_n \in \mathcal{M}_{\alpha,y_n} = \arg\min_{x \in X} J_{\alpha,y_n}(x)$$

satisfies $x_n \rightharpoonup^* \mathcal{M}_{\alpha,y^\delta}$ as defined in (2.7).

Proof. Let

$$f_q(a) = a^q,$$

then from Lemma 1.29 it follows that for all $q \geq 1$ and $a, b \geq 0$

$$(a+b)^q = f_q(a+b) \leq \frac{f_q(2a)}{2} + \frac{f_q(2b)}{2} = 2^{q-1}(a^q + b^q) \quad (3.1)$$

holds. According to Conditions 2.4 and 2.7

$$0 \in \mathcal{D} = \mathrm{dom}(F) \cap \mathrm{dom}(\Psi)$$

and using (3.1) we obtain

$$\Psi(x_n) \leq \frac{J_{\alpha,y_n}(0)}{\alpha} \leq \frac{2^{q-1}}{\alpha} \left(\left\| F(0) - y^\delta \right\|^q + \left\| y^\delta - y_n \right\|^q \right).$$

Since $y_n \to y^\delta$ there exists $N \in \mathbb{N}$ such that

$$\Psi(x_n) \leq C := 2^{q-1} \left(\left\| F(0) - y^\delta \right\|^q + 1 \right) / \alpha$$

whenever $n > N$. Due to the weak coercivity of Ψ we obtain that the sequence $\{x_n\}_{n>N}$ is bounded and admits a weak-* convergent subsequence $\{x_k\} \subset \{x_n\}_{n>N}$. Using the weak-* l.s.c. of G in Lemma 2.6 and Ψ in Condition 2.7 together with $y_n \to y^\delta$ we get for the limit point $\bar{x}$ of $\{x_k\}$

$$\begin{aligned} J_{\alpha,y^\delta}(\bar{x}) &\leq \liminf_{k \to \infty} \left\{ \left\| F(x_k) - y_k \right\|^q + \alpha \Psi(x_k) \right\} \\ &\leq \limsup_{k \to \infty} \left\{ \left\| F(x_k) - y_k \right\|^q + \alpha \Psi(x_k) \right\} \\ &\leq \lim_{k \to \infty} \left\{ \left\| F(x) - y_k \right\|^q + \alpha \Psi(x) \right\} && \forall x \in X \\ &= \left\| F(x) - y^\delta \right\|^q + \alpha \Psi(x) && \forall x \in X, \end{aligned}$$

which proves that $\bar{x} \in \mathcal{M}_{\alpha,y^\delta}$.

The same argument now applies to any subsequence of the original sequence $\{x_n\}$ and yields another subsequence that converges weak-* to $\mathcal{M}_{\alpha,y^\delta}$. According to Lemma 2.10 therefore $x_n \rightharpoonup^* \mathcal{M}_{\alpha,y^\delta}$. □

3.2 Convergence

We have already seen in Section 3.1 that the regularized problem of minimizing (2.5) has a solution and that the solutions depends on the data in a stable way. Therefore, the Tikhonov regularization $\{R_\alpha\}$ fulfills properties (i) and (iii) of Definition 2.1.

In this section we show that the minimizers of the Tikhonov functional converge to the set of Ψ-minimizing solutions. As indicated in the formulation of Definition 2.1 (iv), convergence can only take place if the parameter α is suitably chosen in dependence of the noise level δ and/or the noisy data y^δ. For a-priori parameter choice rules, where $\alpha = \alpha(\delta)$ only, results on the convergence of Tikhonov regularization can be found, for example, in [17, 56] and for a noise-free parameter choice rules, where $\alpha = \alpha(y^\delta)$, e.g., in [39]. Here, we will show convergence when choosing the regularization parameter α according to an a-posteriori parameter choice rule, i.e., $\alpha = \alpha(\delta, y^\delta)$, namely, Morozov's discrepancy principle (see [1, 2, 6]). It is defined as follows.

Definition 3.3. (MDP) Let $\tau_2 \geq \tau_1 \geq 1$ and $y \in \mathrm{rg}(F)$ be fixed. For $\delta > 0$ and $y^\delta \in Y$ with $\|y - y^\delta\| \leq \delta$ we say that the regularization parameter $\alpha = \alpha(\delta, y^\delta) > 0$ is chosen according to *Morozov's Discrepancy Principle (MDP)* if there exists $x_\alpha^\delta \in \mathcal{M}_{\alpha, y^\delta}$ such that

$$\tau_1 \delta \leq \left\| F(x_\alpha^\delta) - y^\delta \right\| \leq \tau_2 \delta. \tag{3.2}$$

Remark 3.4. As mentioned earlier, one of the problems that arise from Tikhonov regularization is that the minimizers $x_\alpha^\delta \in \mathcal{M}_{\alpha, y^\delta}$ are not unique. Through the use of MDP, where we are only interested in those x_α^δ that satisfy (3.2), the set of acceptable regularized solutions may be narrowed down. But not only is it in general still not possible to show uniqueness, we even need to reconsider the question of existence. One easily constructs examples, where a parameter α with $x_\alpha^\delta \in \mathcal{M}_{\alpha, y^\delta}$ fulfilling (3.2) does not exist. This issue will be addressed in depth in Chapter 5, where we formulate a sufficient condition for the existence of such α and also study an example where – depending on y and y^δ – there may or may not exist α satisfying MDP. For the time being, we will, however, simply assume that to every δ and y^δ at least one such $\alpha = \alpha(\delta, y^\delta)$ exists. ∎

In [62, Section 2.5] Tikhonov et. al. show that the following conditions ensure weak convergence of the regularized solutions to the set $\mathcal{L}$ of Ψ-minimizing solutions. The proof is repeated here for the convenience of the reader in the special case that is of interest to us.

Lemma 3.5. *Assume that F, Ψ satisfy Conditions 2.4 and 2.7, respectively. Let $y \in \mathrm{rg}(F)$ be arbitrary but fixed, $\{\delta_n\}_{n \in \mathbb{N}} \subset \mathbb{R}$ be such that $\delta_n \to 0$ and y_n be a sequence in Y with $\|y - y_n\| \leq \delta_n$ for all $n \in \mathbb{N}$. If $\{x_n\} \subset \mathcal{D} = \mathrm{dom}(F) \cap \mathrm{dom}(\Psi)$ satisfies*

$$\lim_{n \to \infty} \left\| F(x_n) - y^{\delta_n} \right\| = 0 \tag{3.3}$$

$$\limsup_{n \to \infty} \Psi(x_n) \leq \psi^\dagger, \tag{3.4}$$

then $x_n \rightharpoonup^* \mathcal{L}$ and $\Psi(x_n) \to \psi^\dagger$, where $\mathcal{L}, \psi^\dagger$ are as defined in (2.3) and (2.4).

Proof. From (3.4) it is clear that the sequence $\{\Psi(x_n)\}$ is bounded. Thus, the same holds true for $\{x_n\}$ due to the weak coercivity of Ψ and we can extract a subsequence $x_{n'} \rightharpoonup^* \bar{x}$. Using (3.3) and the weak-* l.s.c. of $G(x)$ (cf. Lemma 2.6) we obtain

$$G(\bar{x}) = \|F(\bar{x}) - y\| \le \liminf_{n' \to \infty} \|F(x_{n'}) - y\|$$
$$\le \liminf_{n' \to \infty} \left\{ \left\|F(x_{n'}) - y^{\delta_{n'}}\right\| + \delta_{n'} \right\}$$
$$= 0,$$

and, similarly, from (3.4) and the weak-* l.s.c. of Ψ (cf. Condition 2.7)

$$\Psi(\bar{x}) \le \liminf_{n' \to \infty} \Psi(x_{n'})$$
$$\le \limsup_{n' \to \infty} \Psi(x_{n'})$$
$$\le \Psi(x^\dagger).$$

But $x^\dagger$ was chosen to be a Ψ-minimizing solution and therefore $\Psi(\bar{x}) = \Psi(x^\dagger)$ whence it follows that $\bar{x} \in \mathcal{L}$ and $\Psi(x_{n'}) \to \Psi(x^\dagger)$.

The same reasoning applies to any subsequence of $\{x_n\}$ and yields a subsequence weakly converging to $\mathcal{L}$. Therefore the whole sequence weakly converges to $\mathcal{L}$ and $\Psi(x_n) \to \Psi(x^\dagger)$. □

In the following proof of weak convergence for the regularized solutions found through Morozov's discrepancy principle we use techniques similar to [62]. Similar to the proof of stability in Theorem 3.2 we show convergence for arbitrary cuts (cf. Definition 2.3) of $\{\mathcal{M}_{\alpha(\delta, y^\delta), y^\delta}\}_{\delta > 0}$ satisfying (3.2) to the set $\mathcal{L}$ of Ψ-minimizing solutions.

Theorem 3.6. *Let $F, \Psi, \{\delta_n\}, \{y_n\}$ be as defined in Lemma 3.5. If $\alpha_n = \alpha(\delta_n, y^{\delta_n})$ is chosen according to MDP and $x_n \in \mathcal{M}_{\alpha_n, y_n}$ satisfies (3.2) for all $n \in \mathbb{N}$, then the $x_n \rightharpoonup^* \mathcal{L}$ and $\Psi(x_n) \to \psi^\dagger$.*

Proof. From (3.2) we know that

$$\lim_{n \to \infty} \left\|F(x_n) - y^{\delta_n}\right\| \le \lim_{n \to \infty} \tau_2 \delta_n = 0$$

and also that for $x^\dagger \in \mathcal{L}$ it holds true

$$\tau_1^q \delta_n^q + \alpha_n \Psi(x_n) \le \left\|F(x_n) - y^{\delta_n}\right\|^q + \alpha_n \Psi(x_n) \le \delta_n^q + \alpha_n \Psi(x^\dagger).$$

Therefore, we obtain

$$0 \le (\tau_1^q - 1) \frac{\delta_n^q}{\alpha_n} \le \Psi(x^\dagger) - \Psi(x_n) \qquad (3.5)$$

whence it follows that

$$\limsup_{n \to \infty} \Psi(x_n) \le \Psi(x^\dagger).$$

Altogether we have shown that the family $\{x_n\}$ satisfies the assumptions of Lemma 3.5 which gives the assertion. □

For certain penalty terms one can even show convergence in norm or with respect to the topology induced by the penalty term Ψ. For this to hold the penalty term has to fulfill the generalized *Kadec* properties in Condition 3.7 or 3.9, respectively.

Condition 3.7. Let $\{x_n\}_{n\in\mathbb{N}} \subset X$ be such that $x_n \rightharpoonup^* \bar{x} \in X$ and $\Psi(x_n) \to \Psi(\bar{x}) < \infty$, then x_n converges strongly to $\bar{x}$, i.e.,

$$\|x_n - \bar{x}\| \to 0.$$

Theorem 3.8. *Let $F, \Psi, \{\delta_n\}, \{y_n\}$ be as defined in Lemma 3.5 and let Ψ satisfy Condition 3.7. If $\alpha_n = \alpha(\delta_n, y^{\delta_n})$ is chosen according to MDP and $x_n \in \mathcal{M}_{\alpha_n, y_n}$ satisfies (3.2), then x_n converges to $\mathcal{L}$ in norm.*

Proof. The sequence $\{x_n\}$ satisfies the assumptions of Theorem 3.6 and hence also of Lemma 3.5. From the proof of Lemma 3.5 we see that $\{x_n\}$ has a subsequence $x_{n'} \rightharpoonup^* x^\dagger \in \mathcal{L}$. According to Theorem 3.6 also $\Psi(x_{n'}) \to \Psi(x^\dagger)$ holds and we obtain from Condition 3.7

$$\left\|x_{n'} - x^\dagger\right\| \to 0.$$

The same reasoning applies to any subsequence of $\{x_n\}$ and yields a subsequence converging strongly to $\mathcal{L}$. According to Lemma 2.10 therefore the whole sequence converges to $\mathcal{L}$. □

In the same way we derive convergence with respect to the penalty term Ψ if the following condition holds.

Condition 3.9. Let $\{x_n\} \subset X$ be such that $x_n \rightharpoonup^* \bar{x} \in X$ and $\Psi(x_n) \to \Psi(\bar{x}) < \infty$, then x_n converges to $\bar{x}$ with respect to Ψ, i.e.,

$$\Psi(x_n - \bar{x}) \to 0.$$

Theorem 3.10. *Let $F, \Psi, \{\delta_n\}, \{y_n\}$ be as defined in Lemma 3.5 and let Ψ satisfy Condition 3.9. If $\alpha_n = \alpha(\delta_n, y^{\delta_n})$ is chosen according to MDP and $x_n \in \mathcal{M}_{\alpha_n, y_n}$ satisfies (3.2), then x_n converges to $\mathcal{L}$ with respect to Ψ.*

Proof. We argue as in the proof of Theorem 3.8 only using Condition 3.9 instead of Condition 3.7 which gives

$$\Psi(x_{n'} - x^\dagger) \to 0$$

first for subsequences and then from Lemma 2.10 we see that the whole sequence Ψ-converges to $\mathcal{L}$. □

Remark 3.11. If a functional $\Psi(x)$ is coercive in the sense that

$$\|x\|^r \leq c\Psi(x)$$

holds for some $c, r > 0$ and all $x \in X$ then Condition 3.9 clearly implies Condition 3.7.

It has been shown in [24, Lemma 2] that the weighted ℓ_p-norms $\Psi_{p,w}(x)$ in Definition 2.17 satisfy Condition 3.9 and due to Lemma 2.20 they therefore also satisfy Condition 3.7. Note that the same then clearly holds for $\Psi_{p,w}(x)^{1/p}$. ∎

We have thus shown that Tikhonov regularization with regularization parameter chosen according to Morozov's discrepancy principle constitutes a (weak-*) convergent regularization method in the sense of Definition 2.1. Note that in order to show the convergence of this method it was not necessary to use $\alpha(\delta, y^\delta) \to 0$ as $\delta \to 0$. In fact, this may not even be the case for MDP. However, we will see in Section 5.2 that α does tend to zero, for example, if $F(x)$ is continuously differentiable in $x^\dagger$. We summarize our findings in the concluding Theorem.

Theorem 3.12. *The family of Tikhonov operators* $\{R_\alpha\}_{\alpha>0}$ *in Definition 2.1 with regularization parameter* $\alpha = \alpha(\delta, y^\delta)$ *chosen according to MDP in Definition 3.3 is a convergent regularization method for* $F^\dagger$ *as in Definition 2.11. This is only true, however, if we dispense with uniqueness of the regularized solutions and use a set-valued definition of convergence. Here we assume that an parameter* α *satisfying MDP exists for all* $\delta > 0$ *and* y^δ *such that* $\|y - y^\delta\| \leq \delta$.

Chapter 4
The discrepancy functionals

We now turn our attention to the question of existence of an parameter $\alpha = \alpha(\delta, y^\delta)$ according to Morozov's Discrepancy Principle (MDP) for Tikhonov regularization. In the case of linear operators F it has been shown in [6] that existence can be guaranteed, for example, if the penalty term $\Psi(x)$ is strictly convex on the nullspace $N(F)$. Related questions have been studied earlier also in [45, 61, 62]. For non-linear operators F and the special choice $\Psi(x) = \|x\|^2$ in Hilbert spaces – we refer to this combination as *classical Tikhonov regularization* – a sufficient condition for the existence of α according to MDP has been formulated in [47] and this condition has been generalized to convex penalty functionals satisfying Condtion 2.7 in [1].

To analyze when it is possible to find α that admits $x_\alpha^\delta \in \mathcal{M}_{\alpha,y^\delta}$ satisfying

$$\tau_1 \delta \leq \left\| F(x_\alpha^\delta) - y^\delta \right\| \leq \tau_2 \delta. \tag{4.1}$$

we study the following (set-valued) functionals, which we will refer to as *discrepancy functionals*. Recall that for any set A we denote its power set by 2^A.

Definition 4.1. Let $y \in \mathrm{rg}(F)$, $\delta > 0$ and $y^\delta \in Y$ with $\|y - y^\delta\| \leq \delta$ be fixed. We introduce the shorthand notation

$$G(x) = \left\| F(x) - y^\delta \right\|. \tag{4.2}$$

Then, for $\alpha \in (0, \infty)$ we define the *discrepancy functionals* $g, w : \mathbb{R}^+ \to 2^{\mathbb{R}_0^+}$ and $m : \mathbb{R}^+ \to \mathbb{R}_0^+$ through

$$g(\alpha) = \left\{ G(x_\alpha^\delta) \mid x_\alpha^\delta \in \mathcal{M}_{\alpha,y^\delta} \right\}, \tag{4.3}$$

$$w(\alpha) = \left\{ \Psi(x_\alpha^\delta) \mid x_\alpha^\delta \in \mathcal{M}_{\alpha,y^\delta} \right\}, \tag{4.4}$$

$$m(\alpha) = J_{\alpha,y^\delta}(x_\alpha^\delta), \qquad x_\alpha^\delta \in \mathcal{M}_{\alpha,y^\delta}. \tag{4.5}$$

Remark 4.2. Even if there exist multiple minimizers $x_\alpha^\delta \in \mathcal{M}_{\alpha,y^\delta}$, they all share the same value of $J_{\alpha,y^\delta}(x_\alpha^\delta)$ as otherwise not all of them would be (global) minimizers. Thus the value of $m(\alpha)$ does not depend on the particular choice of $x_\alpha^\delta \in \mathcal{M}_{\alpha,y^\delta}$. This, however, is in general not true for $G(x_\alpha^\delta)$ and $\Psi(x_\alpha^\delta)$. The discrepancy functionals are clearly linked through the Tikhonov functional $J_{\alpha,y^\delta}(x)$ in (2.5). We have

$$g(\alpha) = (m(\alpha) - \alpha w(\alpha))^{1/q} = \left\{ (m(\alpha) - \alpha \omega)^{1/q} \mid \omega \in w(\alpha) \right\}. \qquad (4.6)$$

■

Let us start by repeating the definition of Morozov's discrepancy principle (compare to Definition 3.3) with the use of the discrepancy functional $g(\alpha)$ defined in (4.3).

Definition 4.3. Let $\tau_2 \geq \tau_1 \geq 1$ and $y \in \mathrm{rg}(F)$ be fixed. For $\delta > 0$ and $y^\delta \in Y$ with $\|y - y^\delta\| \leq \delta$ we say that the regularization parameter $\alpha = \alpha(\delta, y^\delta) > 0$ is chosen according to *Morozov's Discrepancy Principle (MDP)* if

$$g(\alpha) \cap [\tau_1 \delta, \tau_2 \delta] \neq \emptyset. \qquad (4.7)$$

An immediate consequence of (4.1) which we will need repeatedly, is the following bound on the distance between the regularized solutions and the Ψ-minimizing solutions when measured in the image space.

Lemma 4.4. *Let y, δ and y^δ be as in Definition 4.3. If $\alpha = \alpha(\delta, y^\delta)$ is chosen according to MDP and $x_\alpha^\delta \in \mathcal{M}_{\alpha,y^\delta}$ satisfies (4.1), then for any Ψ-minimizing solution $x^\dagger \in \mathcal{L}$ as defined in (2.3) we have*

$$\left\| F(x_\alpha^\delta) - F(x^\dagger) \right\| \leq (\tau_2 + 1)\delta. \qquad (4.8)$$

Proof. Using that $F(x^\dagger) = y$ for all $x^\dagger \in \mathcal{L}$ and that the noisy data satisfy $\|y - y^\delta\| \leq \delta$, we obtain from (4.1)

$$\left\| F(x_\alpha^\delta) - F(x^\dagger) \right\| \leq \left\| F(x_\alpha^\delta) - y^\delta \right\| + \left\| y^\delta - y \right\| \leq (\tau_2 + 1)\delta,$$

which readily proves the assertion of the Lemma. □

The next Lemma will play a key role in our analysis.

Lemma 4.5. *If α is chosen according to MDP, then*

$$\Psi(x_\alpha^\delta) \leq \psi^\dagger$$

holds for all $x_\alpha^\delta \in \mathcal{M}_{\alpha,y^\delta}$ satisfying (4.1), where $\psi^\dagger$ is as defined in (2.4).

Proof. Using (4.1) and the minimizing property of $x_\alpha^\delta \in \mathcal{M}_{\alpha,y^\delta}$ we see that for any $x^\dagger \in \mathcal{L}$

$$\tau_1^q \delta^q + \alpha \Psi(x_\alpha^\delta) \leq \left\| F(x_\alpha^\delta) - y^\delta \right\|^q + \alpha \Psi(x_\alpha^\delta)$$
$$\leq \left\| F(x^\dagger) - y^\delta \right\|^q + \alpha \Psi(x^\dagger)$$
$$\leq \left\| y - y^\delta \right\|^q + \alpha \Psi(x^\dagger)$$
$$\leq \delta^q + \alpha \Psi(x^\dagger)$$

holds. Since $\tau_1 \geq 1$ we thus get

$$0 \leq (\tau_1^q - 1)\frac{\delta^q}{\alpha} \leq \Psi(x^\dagger) - \Psi(x_\alpha^\delta),$$

which completes the proof. □

In the sequel we will consider limits of sequences $g(\alpha_n)$ or $w(\alpha_n)$ as $\alpha_n \to \alpha$. They are to be understood as limits of cuts of the set-valued functions g, w (compare to Definition 2.3). We repeat the definition of such cuts in this context.

Definition 4.6. If $\{\alpha_n\}_{n \in \mathbb{N}}$ is a positive sequence and $f(\alpha)$ is a setvalued function (such as g and w in (4.3) and (4.4) above), then a *cut* of the sequence $\{f(\alpha_n)\}_{n \in \mathbb{N}}$ of sets is defined to be a sequence $\{f_n\}_{n \in \mathbb{N}}$ of arbitrary elements $f_n \in f(\alpha_n)$ for all $n \in \mathbb{N}$.

4.1 Monotonicity

In the following we collect basic properties of the functionals g, w and m from [62, Section 2.6].

Lemma 4.7. *The functional $w(\alpha)$ is non-increasing and the functionals $g(\alpha), m(\alpha)$ are non-decreasing for $\alpha \in (0, \infty)$ in the sense that if $0 < \alpha < \beta$ then*

$$\sup g(\alpha) \leq \inf g(\beta),$$
$$\inf w(\alpha) \geq \sup w(\beta),$$
$$m(\alpha) \leq m(\beta).$$

Proof. Let $0 < \alpha < \beta$ and choose $x_\alpha^\delta \in \mathcal{M}_{\alpha,y^\delta}$, $x_\beta^\delta \in \mathcal{M}_{\beta,y^\delta}$ arbitrary but fixed, then

$$G(x_\alpha^\delta)^q + \alpha \Psi(x_\alpha^\delta) \leq G(x_\beta^\delta)^q + \alpha \Psi(x_\beta^\delta)$$
$$G(x_\beta^\delta)^q + \beta \Psi(x_\beta^\delta) \leq G(x_\alpha^\delta)^q + \beta \Psi(x_\alpha^\delta).$$

Combining these inequalities we get
$$\frac{1}{\alpha}\left(G(x_\alpha^\delta)^q - G(x_\beta^\delta)^q\right) \leq \Psi(x_\beta^\delta) - \Psi(x_\alpha^\delta) \leq \frac{1}{\beta}\left(G(x_\alpha^\delta)^q - G(x_\beta^\delta)^q\right).$$
Since $0 < 1/\beta < 1/\alpha$ we may conclude that $G(x_\alpha^\delta)^q - G(x_\beta^\delta)^q \leq 0$ and obtain
$$G(x_\alpha^\delta) \leq G(x_\beta^\delta)$$
$$\Psi(x_\beta^\delta) \leq \Psi(x_\alpha^\delta).$$
Since x_α^δ and x_β^δ were arbitrary elements in $\mathcal{M}_{\alpha,y^\delta}$ and $\mathcal{M}_{\beta,y^\delta}$, respectively, it must then hold that $\sup g(\alpha) \leq \inf g(\beta)$ and $\inf w(\alpha) \geq \sup w(\beta)$. Finally, the estimates
$$\begin{aligned}m(\alpha) &= G(x_\alpha^\delta)^q + \alpha\Psi(x_\alpha^\delta) \\ &\leq G(x_\beta^\delta)^q + \alpha\Psi(x_\beta^\delta) \\ &\leq G(x_\beta^\delta)^q + \beta\Psi(x_\beta^\delta) = m(\beta)\end{aligned}$$
complete the proof. □

It is an interesting consequence of the monotonicity of the discrepancy functionals g and w, that multiple global minimizers x_α^δ of $J_{\alpha,y^\delta}(x)$ with different values of $G(x_\alpha^\delta)$ and $\Psi(x_\alpha^\delta)$ can only exist for countably many values of α. This result has been shown in [36] for linear operators.

Lemma 4.8. *The set*
$$A = \left\{\alpha > 0 \mid \inf g(\alpha) < \sup g(\alpha)\right\}$$
is at most countable and the functional $g(\alpha)$ is single-valued everywhere else. The same holds true for $w(\alpha)$ and the respective sets coincide.

Proof. We first show that the set
$$B = \left\{\beta > 0 \mid \inf w(\beta) < \sup w(\beta)\right\}$$
coincides with A. If $\beta \notin B$ and ω is the unique value in $w(\beta)$, then for any two minimizers $x_1, x_2 \in \mathcal{M}_{\beta,y^\delta}$ it holds that
$$G(x_1)^q + \beta\omega = G(x_2)^q + \beta\omega.$$
Therefore $g(\beta)$ is single-valued and hence $\beta \notin A$. Reversing the argument we also see that if $g(\alpha)$ is single-valued then so is $w(\alpha)$ and if follows that $A = B$.

To show that the sets are at most countable in size, we note that for every value $\alpha \in A$ the set $g(\alpha)$ contains more than one element. Consequently, the open interval
$$(\inf g(\alpha), \sup g(\alpha))$$

is nonempty and hence contains a rational number. Due to the monotonicity of g which was established in the Lemma 4.7, these intervals are disjoint for different values α and since there are only countably many rationals there can be no more than countably many elements in A. □

Lemma 4.9. *The values of $g(\alpha)$ and $w(\alpha)$ are bounded by*

$$\sup g(\alpha) \leq m(\alpha)^{1/q}, \quad \sup w(\alpha) \leq \frac{m(\alpha)}{\alpha}$$

for each $\alpha > 0$ and $m(\alpha)$ is uniformely bounded by

$$m(\alpha) \leq J_{\alpha,y^\delta}(0) = \left\|F(0) - y^\delta\right\|^q < \infty. \tag{4.9}$$

Proof. Note that according to Conditions 2.4 and 2.7

$$0 \in \mathcal{D} = \mathrm{dom}(F) \cap \mathrm{dom}(\Psi)$$

and therefore $J_{\alpha,y^\delta}(0)$ is finite. For any $x_\alpha^\delta \in \mathcal{M}_{\alpha,y^\delta}$ it holds that

$$G(x_\alpha^\delta)^q \leq G(x_\alpha^\delta)^q + \alpha\Psi(x_\alpha^\delta) = m(\alpha)$$

and

$$\alpha\Psi(x_\alpha^\delta) \leq G(x_\alpha^\delta)^q + \alpha\Psi(x_\alpha^\delta) = m(\alpha),$$

which establishes the upper bounds in terms of $m(\alpha)$. The uniform bound for $m(\alpha)$ comes from

$$m(\alpha) = G(x_\alpha^\delta)^q + \alpha\Psi(x_\alpha^\delta) \leq G(0)^q + \alpha\Psi(0) = \left\|F(0) - y^\delta\right\|^q.$$

□

4.2 Continuity

We will now show that the functional $m(\alpha)$ is continuous. Similar results have been proven for linear operators and under slightly different assumptions on the penalty term in [6, 36], and in [47] for non-linear operators F and classical Tikhonov regularization, i.e., the specific choice $\Psi(x) = \|x\|^2$. The continuity is based on the following convergence result for the minimizers of the Tikhonov functional.

Proposition 4.10. *Let $\{\alpha_n\}$ be a positive sequence with $\alpha_n \to \bar{\alpha} > 0$ and let $\{x_n\}_{n\in\mathbb{N}}$ be a sequence of corresponding minimizers of $J_{\alpha_n,y^\delta}(x)$. Then*

$$x_n \rightharpoonup^* \mathcal{M}_{\bar{\alpha},y^\delta}.$$

Proof. There exists $\varepsilon > 0$ such that $\alpha_n \geq \varepsilon$ for all n. Using the monotonicity of $w(\alpha)$ in Lemma 4.7 and the boundedness in Lemma 4.9 we thus obtain that the sequence $\{\Psi(x_n)\}$ is uniformly bounded by

$$\Psi(x_n) \leq w(\varepsilon) \leq \frac{m(\varepsilon)}{\varepsilon} \leq \frac{1}{\varepsilon} \left\| F(0) - y^\delta \right\|^q < \infty$$

and due to the weak coercivity of Ψ, so is $\{x_n\}$. Therefore we can extract a subsequence $\{x_{n'}\}$ with $x_{n'} \rightharpoonup^* \bar{x}$. We now show that $\bar{x}$ is indeed a minimizer of $J_{\bar{\alpha},y^\delta}(.)$. To this end we note that by the weak-* lower semicontinuity of $\left\| F(.) - y^\delta \right\|$ and Ψ from Lemma 2.6 and Condition 2.7, respectively, as well as the minimizing property of the $x_{n'}$, we get

$$\left\| F(\bar{x}) - y^\delta \right\|^q + \bar{\alpha}\Psi(\bar{x}) \leq \liminf_{n' \to \infty} \left\{ \left\| F(x_{n'}) - y^\delta \right\|^q + \alpha_{n'}\Psi(x_{n'}) \right\}$$
$$\leq \limsup_{n' \to \infty} \left\{ \left\| F(x_{n'}) - y^\delta \right\|^q + \alpha_{n'}\Psi(x_{n'}) \right\}$$
$$\leq \lim_{n' \to \infty} \left\{ \left\| F(x) - y^\delta \right\|^q + \alpha_{n'}\Psi(x) \right\} \quad \forall x \in X$$
$$= \left\| F(x) - y^\delta \right\|^q + \bar{\alpha}\Psi(x) \quad \forall x \in X,$$

which readily proves that $\bar{x}$ is a minimizer of $J_{\bar{\alpha},y^\delta}(.)$.

This argument applies to every subsequence of the original sequence $\{x_n\}$ and yields a subsequence which converges weak-* to a minimizer of the Tikhonov functional $J_{\bar{\alpha},y^\delta}(.)$. Therefore the entire sequence $\{x_n\}$ converges in a weak-* sense to the set $\mathcal{M}_{\bar{\alpha},y^\delta}$ of all such minimizers according to Lemma 2.10. □

On the basis of Proposition 4.10 we now show that $m(\alpha)$ is continuous.

Lemma 4.11. *The functional $m(\alpha)$ defined in (4.5) is continuous on $(0,\infty)$.*

Proof. Let $\alpha_n \to \bar{\alpha} > 0$ be a positive sequence and let $x_n \in \mathcal{M}_{\alpha_n,y^\delta}$. It immediately follows that for any $\bar{x} \in \mathcal{M}_{\bar{\alpha},y^\delta}$

$$\limsup_{n \to \infty} m(\alpha_n) = \limsup_{n \to \infty} \left\{ \left\| F(x_n) - y^\delta \right\|^q + \alpha_n \Psi(x_n) \right\}$$
$$\leq \limsup_{n \to \infty} \left\{ \left\| F(\bar{x}) - y^\delta \right\|^q + \alpha_n \Psi(\bar{x}) \right\} \quad (4.10)$$
$$= m(\bar{\alpha}).$$

To obtain the converse estimate we note that Proposition 4.10 states that $x_n \rightharpoonup^* \mathcal{M}_{\bar{\alpha},y^\delta}$. Therefore any subsequence of $\{x_n\}$ contains a subsequence $\{x_k\}$ such that $x_k \rightharpoonup^* \bar{x}$ for some $\bar{x} \in \mathcal{M}_{\bar{\alpha},y^\delta}$ and the weak-* lower semicontinuity of G and Ψ yields

$$m(\bar{\alpha}) = \left\|F(\bar{x}) - y^\delta\right\|^q + \bar{\alpha}\Psi(\bar{x})$$
$$\leq \liminf_{k\to\infty}\left\{\left\|F(x_k) - y^\delta\right\|^q + \alpha_k\Psi(x_k)\right\}$$
$$= \liminf_{k\to\infty} m(\alpha_k).$$

Together with (4.10) this shows that $m(\alpha_k) \to m(\bar{\alpha})$. Using the convergence principle described in Lemma 2.9 we obtain that $m(\alpha_n) \to m(\bar{\alpha})$ for the whole sequence α_n. □

As mentioned earlier, the values of $G(x_\alpha^\delta)$ in $g(\alpha)$ will in general depend on the choice of $x_\alpha^\delta \in \mathcal{M}_{\alpha,y^\delta}$. Next we prove, however, that g is left and right continuous with respect to $\alpha \in (0,\infty)$ (compare [6, 61] for the linear and [47] for the non-linear classical Tikhonov case). Consequently, g is continuous at α whenever $G(x)$ is constant on $\mathcal{M}_{\alpha,y^\delta}$ (cf. Corollary 4.14).

Lemma 4.12. *To each $\bar{\alpha} > 0$ there exist $\bar{x}_1, \bar{x}_2 \in \mathcal{M}_{\bar{\alpha},y^\delta}$ such that*

$$G(\bar{x}_1) = \inf g(\bar{\alpha}),$$
$$G(\bar{x}_2) = \sup g(\bar{\alpha}).$$

and it holds that

$$\lim_{\alpha\to\bar{\alpha}^-} g(\alpha) = \inf g(\bar{\alpha}),$$
$$\lim_{\alpha\to\bar{\alpha}^+} g(\alpha) = \sup g(\bar{\alpha}).$$

Here, the limits for $g(\alpha)$ are understood as limits of arbitrary cuts of these set-valued functions as in Definition 4.6.

Proof. Let $\{\alpha_n\}$ be a positive, strictly increasing sequence converging to $\bar{\alpha}$, and $\{x_n\}$ be a corresponding sequence of arbitrary minimizers of the Tikhonov functionals $J_{\alpha_n,y^\delta}(x)$, i.e., $x_n \in \mathcal{M}_{\alpha_n,y^\delta}$. From Proposition 4.10 we obtain that $\{x_n\}$ contains a subsequence, denoted by $\{x_{n'}\}$, such that $x_{n'} \rightharpoonup^* \bar{x} \in \mathcal{M}_{\bar{\alpha},y^\delta}$. It then follows from the weak-* sequential lower semi-continuity of $G(x)$ in Lemma 2.6 and the monotonicity of $g(\alpha)$ in Lemma 4.7 that

$$G(\bar{x}) \leq \liminf_{n'\to\infty} G(x_{n'})$$
$$\leq \limsup_{n'\to\infty} G(x_{n'})$$
$$\leq \inf_{x\in\mathcal{M}_{\bar{\alpha},y^\delta}} G(x)$$
$$\leq G(\bar{x})$$

where the last inequatlity holds since $\bar{x} \in \mathcal{M}_{\bar{\alpha},y^\delta}$. Together this implies that

$$G(x_{n'}) \to G(\bar{x}) = \inf_{x \in \mathcal{M}_{\bar{\alpha},y^\delta}} G(x)$$

and we have found an element $\bar{x}_1 := \bar{x} \in \mathcal{M}_{\bar{\alpha},y^\delta}$ such that $G(\bar{x}_1) = \inf g(\bar{\alpha})$. The same reasoning can be applied to any subsequence of the original sequence $\{x_n\}$ (i.e., any subsequence of $\{G(x_n)\}$), which – according to Lemma 2.9 – shows that

$$G(x_n) \to \inf_{x \in \mathcal{M}_{\bar{\alpha},y^\delta}} G(x). \qquad (4.11)$$

In the case where $\{\alpha_n\}$ is a strictly decreasing sequence with limit $\bar{\alpha} > 0$, we will use the same argument, but applied to w. As above we may choose a subsequence $\{x_{n'}\}$ of a sequence of arbitrary minimizers $\{x_n\}$ such that $x_{n'} \rightharpoonup^* \bar{x} \in \mathcal{M}_{\bar{\alpha},y^\delta}$. Now, the weak-* lower semicontinuity of Ψ and the monotonicity of $w(\alpha)$ (cf. Lemma 4.7) imply that

$$\Psi(\bar{x}) \leq \liminf_{n' \to \infty} \Psi(x_{n'})$$
$$\leq \limsup_{n' \to \infty} \Psi(x_{n'})$$
$$\leq \inf w(\bar{\alpha})$$
$$\leq \Psi(\bar{x}),$$

since again $\bar{x} \in \mathcal{M}_{\bar{\alpha},y^\delta}$, which means that

$$\Psi(x_{n'}) \to \Psi(\bar{x}) = \inf_{x \in \mathcal{M}_{\bar{\alpha},y^\delta}} \Psi(x).$$

Again these arguments can be applied to any subsequence of $\{x_n\}$ (and therefore $\{\Psi(x_n)\}$), and the convergence principle in Lemma 2.9 yields that

$$\Psi(x_n) \to \Psi(\bar{x}) = \inf_{x \in \mathcal{M}_{\bar{\alpha},y^\delta}} \Psi(x). \qquad (4.12)$$

Using the continuity of $m(\alpha)$ from Lemma 4.11 we obtain from (4.12) that

$$\lim_{n \to \infty} G(x_n) = \lim_{n \to \infty} \left(m(\alpha_n) - \alpha_n \Psi(x_n) \right)^{1/q}$$
$$= \left(m(\bar{\alpha}) - \bar{\alpha} \Psi(\bar{x}) \right)^{1/q}$$
$$= G(\bar{x}).$$

Moreover, it is an immediate consequence of (4.12) and $m(\alpha)$ being single valued that for any $x \in \mathcal{M}_{\bar{\alpha},y^\delta}$ we have

$$G(\bar{x})^q = m(\bar{\alpha}) - \bar{\alpha}\Psi(\bar{x}) \geq m(\bar{\alpha}) - \bar{\alpha}\Psi(x) = G(x)^q.$$

Therefore
$$\lim_{n\to\infty} G(x_n) = G(\bar{x}) = \sup_{x\in\mathcal{M}_{\bar{\alpha},y^\delta}} G(x). \tag{4.13}$$
and we have constructed $\bar{x}_2 := \bar{x} \in \mathcal{M}_{\bar{\alpha},y^\delta}$ such that $G(\bar{x}_2) = \sup g(\bar{\alpha})$.

The minimizers $x_n \in \mathcal{M}_{\alpha_n,y^\delta}$ corresponding to $\{\alpha_n\}$ were chosen arbitrarily and each sequence $\{G(x_n)\}$ represents one of the cuts of $\{g(\alpha_n)\}$. Thus (4.11) and (4.13), respectively, yield the asymptotical relations
$$\lim_{\alpha\to\bar{\alpha}^-} g(\alpha) = \inf g(\bar{\alpha}),$$
$$\lim_{\alpha\to\bar{\alpha}^+} g(\alpha) = \sup g(\bar{\alpha}).$$
□

Corollary 4.13. *To each $\bar{\alpha} > 0$ there exist $\bar{x}_1, \bar{x}_2 \in \mathcal{M}_{\bar{\alpha},y^\delta}$ such that*
$$\Psi(\bar{x}_1) = \sup w(\bar{\alpha}),$$
$$\Psi(\bar{x}_2) = \inf w(\bar{\alpha}).$$
and it holds that
$$\lim_{\alpha\to\bar{\alpha}^+} w(\alpha) = \sup w(\bar{\alpha}),$$
$$\lim_{\alpha\to\bar{\alpha}^-} w(\alpha) = \inf w(\bar{\alpha}).$$
Here, the limits for $w(\alpha)$ are understood as in Definition 4.6.

Proof. In Lemma 4.12 we have seen that to each $\alpha > 0$ we can find $x_1, x_2 \in \mathcal{M}_{\alpha,y^\delta}$ such that
$$\left\|F(x_1) - y^\delta\right\| = \inf g(\alpha)$$
$$\left\|F(x_2) - y^\delta\right\| = \sup g(\alpha).$$
Note, that because the value of the Tikhonov functional $m(\alpha)$ is constant for all minimizers, the same elements x_1, x_2 also produce the maximal and minimal value of $\Psi(x)$ in $\mathcal{M}_{\alpha,y^\delta}$, respectively, and that also the reversed left and right continuity results hold, i.e.,
$$\lim_{\alpha\to\bar{\alpha}^-} w(\alpha) = \sup w(\alpha) = \Psi(x_1)$$
$$\lim_{\alpha\to\bar{\alpha}^+} w(\alpha) = \inf w(\alpha) = \Psi(x_2)$$
□

If $G(x_\alpha^\delta)$ takes the same value for all $x_\alpha^\delta \in \mathcal{M}_{\alpha,y^\delta}$ (e.g., if $\mathcal{M}_{\alpha,y^\delta}$ consists of only one element), then it follows from Lemma 4.12 that g is continuous at α. We have thus proven the following corollary.

Corollary 4.14. *If for $\alpha > 0$ the functional G is constant on $\mathcal{M}_{\alpha,y^\delta}$, then g is continuous at α. In this case also Ψ has to be constant on $\mathcal{M}_{\alpha,y^\delta}$ and thus w is continuous as well.*

4.3 Asymptotic behaviour

Lemma 4.15. *The asymptotic values of $g(\alpha), w(\alpha)$ and $m(\alpha)$ are given by*

$$\lim_{\alpha \to 0^+} m(\alpha) = \inf_{x \in X} G(x)^q \qquad \lim_{\alpha \to \infty} m(\alpha) = G(0)^q$$
$$\lim_{\alpha \to 0^+} g(\alpha) = \inf_{x \in X} G(x) \qquad \lim_{\alpha \to \infty} g(\alpha) = G(0)$$
$$\lim_{\alpha \to 0^+} \alpha w(\alpha) = 0 \qquad \lim_{\alpha \to \infty} \alpha w(\alpha) = 0$$

Here, the limits for $g(\alpha)$ and $w(\alpha)$ are understood as the limits of arbitrary cuts of these set-valued functions as in Definition 4.6.

Proof. Note that according to Lemma 4.7 $m(\alpha)$ is monotone and according to Lemma 4.9 it is also bounded,

$$0 \leq m(\alpha) \leq G(0)^q,$$

so that the limits of $m(\alpha)$ as $\alpha \to 0^+$ and $\alpha \to \infty$ exist. Let now $\alpha_n \to \infty$ and $x_n \in \mathcal{M}_{\alpha_n,y^\delta}$, then by virtue of Lemma 4.9 we have

$$\lim_{n \to \infty} \Psi(x_n) \leq \lim_{n \to \infty} \frac{m(\alpha_n)}{\alpha_n} \leq \lim_{n \to \infty} \frac{\|F(0) - y^\delta\|^q}{\alpha_n} = 0$$

We have established in Lemma 2.8 that $x_n \rightharpoonup^* 0$ whenever $\Psi(x_n) \to 0$ and from the weak-* lower semicontinuity of G in Lemma 2.6 we obtain

$$G(0)^q \leq \liminf_{n \to \infty} G(x_n)^q$$
$$\leq \limsup_{n \to \infty} G(x_n)^q$$
$$\leq \lim_{n \to \infty} m(\alpha_n)$$
$$\leq G(0)^q,$$

which shows that $m(\alpha) \to G(0)^q$ and $g(\alpha) \to G(0)$ as $\alpha \to \infty$. Consequently,

$$\lim_{n \to \infty} \alpha_n \Psi(x_n) = \lim_{n \to \infty} m(\alpha_n) - G(x_n)^q = 0.$$

To derive asymptotics as $\alpha \to 0^+$, we take a sequence $\{z_n\}$ in X such that

$$\lim_{n \to \infty} G(z_n) = \inf_{x \in X} G(x).$$

Then for every $n \in \mathbb{N}$ and $\alpha > 0$ it holds that

$$m(\alpha) \leq G(z_n)^q + \alpha \Psi(z_n)$$

and passing to the limit first as $\alpha \to 0^+$ and then as $n \to \infty$ we find that

$$\lim_{\alpha \to 0} m(\alpha) \leq \lim_{n \to \infty} G(z_n)^q = \inf_{x \in X} G(x)^q. \qquad (4.14)$$

Since for any positive sequence $\alpha_n \to 0$ and $x_n \in \mathcal{M}_{\alpha_n, y^\delta}$

$$\inf_{x \in X} G(x)^q \leq G(x_n)^q \leq m(\alpha_n)$$

holds and since the left hand side is independent of α it follows from (4.14) that

$$\inf_{x \in X} G(x)^q \leq \liminf_{n \to \infty} G(x_n)^q$$
$$\leq \limsup_{n \to \infty} G(x_n)^q$$
$$\leq \lim_{n \to \infty} m(\alpha_n)$$
$$\leq \inf_{x \in X} G(x)^q.$$

Therefore $m(\alpha) \to \inf_{x \in X} G(x)^q$ and $g(\alpha) \to \inf_{x \in X} G(x)$ as $\alpha \to 0^+$ so that

$$\lim_{n \to \infty} \alpha_n \Psi(x_n) = \lim_{n \to \infty} m(\alpha_n) - G(x_n)^q = 0.$$

completes the proof. □

4.4 Strict monotonicity

Combining the monotonicity results in Lemma 4.7 with Lemma 4.11 we are now in position to prove a (partial) strict monotonicity result for $m(\alpha) = J_{\alpha,y^\delta}(x_\alpha^\delta)$. It relies on the following Lemma, which points out the limitations of strict monotonicity in the light of the boundedness of $m(\alpha)$ by $J_{\alpha,y^\delta}(0)$ – a value that is independent of α (cf Lemma 4.9).

Lemma 4.16. *If* $m(\alpha) = m(\beta)$ *for* $0 < \alpha < \beta$, *then* $0 \in \mathcal{M}_{\alpha,y^\delta}$ *and* $\mathcal{M}_{\beta,y^\delta} = \{0\}$.

Proof. Let $x_\alpha^\delta \in \mathcal{M}_{\alpha,y^\delta}$ and $x_\beta^\delta \in \mathcal{M}_{\beta,y^\delta}$, then from their minimizing properties we obtain

$$G(x_\beta^\delta)^q + \beta \Psi(x_\beta^\delta) = G(x_\alpha^\delta)^q + \alpha \Psi(x_\alpha^\delta) \leq G(x_\beta^\delta)^q + \alpha \Psi(x_\beta^\delta).$$

This implies that $(\beta-\alpha)\Psi(x_\beta^\delta) \leq 0$ which amounts to $\Psi(x_\beta^\delta) = 0$. According to Condition 2.7 it must then hold that $x_\beta^\delta = 0$ and since x_β^δ was arbitrary in $\mathcal{M}_{\beta,y^\delta}$, also $\mathcal{M}_{\beta,y^\delta} = \{0\}$.

Having established that $x_\beta^\delta = 0$ is the only minimizer of $J_{\beta,y^\delta}(x)$, it follows from $\Psi(0) = 0$ that

$$J_{\alpha,y^\delta}(0) = G(0)^q = m(\beta) = m(\alpha).$$

Therefore $x_\beta^\delta = 0$ is also a minimizer of $J_{\alpha,y^\delta}(x)$ so that $0 \in \mathcal{M}_{\alpha,y^\delta}$. □

Lemma 4.17. *If there exists $\alpha_0 > 0$ such that $0 \in \mathcal{M}_{\alpha_0,y^\delta}$, then the value α^* defined by*

$$\alpha^* = \inf\left\{\alpha > 0 \mid 0 \in \mathcal{M}_{\alpha,y^\delta}\right\} \tag{4.15}$$

is finite and it holds that $0 \in \mathcal{M}_{\alpha^*,y^\delta}$ and $\mathcal{M}_{\alpha,y^\delta} = \{0\}$ for all $\alpha > \alpha^*$. Moreover, if $G(0) > \delta$ then α^* is positive.

Proof. Clearly, $\alpha^* \leq \alpha_0 < \infty$ so that α^* is finite. Let now $\{\alpha_n\}$ be a nonincreasing sequence such that $0 \in \mathcal{M}_{\alpha_n,y^\delta}$ (i.e., $m(\alpha_n) = G(0)^q$) for all $n \in \mathbb{N}$, and $\alpha_n \to \alpha^*$ as $n \to \infty$. To any value $\beta > \alpha^*$ we can then find $n \in \mathbb{N}$ such that $\alpha_n < \beta$. Due to the boundedness and monotonicity of $m(\alpha)$ in Lemma 4.9 and Lemma 4.7 it holds that

$$G(0)^q = m(\alpha_n) \leq m(\beta) \leq G(0)^q$$

whence it follows by virtue of Lemma 4.16 that $\mathcal{M}_{\beta,y^\delta} = \{0\}$ and $m(\beta) = G(0)^q$ for all $\beta > \alpha^*$. Using the continuity of $m(\alpha)$ from Lemma 4.11 we obtain that $m(\alpha^*) = G(0)^q$ which implies that $0 \in \mathcal{M}_{\alpha^*,y^\delta}$.

Finally, to see positivity when $G(0) > \delta$ we use that by virtue of Lemma 4.15

$$\lim_{\alpha \to 0^+} m(\alpha) = \inf_{x \in X} G(x)^q \leq G(x^\dagger)^q \leq \delta^q < G(0)^q,$$

so that due to the continuity of $m(\alpha)$ (cf. Lemma 4.11) there exists $\varepsilon > 0$ such that $m(\alpha) < G(0)^q = J_{\alpha,y^\delta}(0)$ for all $0 < \alpha < \varepsilon$ which implies that $\alpha^* \geq \varepsilon > 0$. □

Corollary 4.18. *If α^* in (4.15) is finite, then the functional $m(\alpha)$ is strictly monotonically increasing for $\alpha \in (0, \alpha^*]$ and for $\alpha \geq \alpha^*$ the value of $m(\alpha)$ is constant with*

$$m(\alpha) = \left\|F(0) - y^\delta\right\|^q.$$

Otherwise, if $0 \notin \mathcal{M}_{\alpha,y^\delta}$ for all $\alpha > 0$, then $\alpha^* = +\infty$ and $m(\alpha)$ is strictly monotonically increasing throughout $\mathbb{R}^+$.

Proof. First, let $\alpha^* = +\infty$, i.e., $0 \notin \mathcal{M}_{\alpha,y^\delta}$ for all $\alpha > 0$. We have seen in Lemma 4.7 that $m(\alpha) \leq m(\beta)$ whenever $0 < \alpha < \beta$ and since $0 \notin \mathcal{M}_{\alpha,y^\delta} \cup \mathcal{M}_{\beta,y^\delta}$ Lemma 4.16 implies that $m(\alpha) \neq m(\beta)$, which readily establishes the strict monotonicity.

On the other hand, if $\alpha^* < \infty$, then Lemma 4.17 asserts that $m(\alpha) = G(0)^q$ for all $\alpha \geq \alpha^*$. For $0 < \alpha < \beta < \alpha^*$ we again obtain strict monotonicity from $m(\alpha) \leq m(\beta)$ in Lemma 4.7 and $0 \notin \mathcal{M}_{\alpha,y^\delta} \cup \mathcal{M}_{\beta,y^\delta}$, which according to Lemma 4.16 implies that $m(\alpha) \neq m(\beta)$. □

4.5 Subdifferentiability

In addition to the continuity and strict monotonicity we also show that the negative discrepancy functional

$$\tilde{m}(\alpha) = -m(\alpha) = -J_{\alpha,y^\delta}(x_\alpha^\delta), \qquad x_\alpha^\delta \in \mathcal{M}_{\alpha,y^\delta}$$

admits a subdifferential $\partial \tilde{m}(\alpha)$ and that $\partial \tilde{m}(\alpha)$ can be expressed in terms of the elements of $w(\alpha)$. This has been shown under the additional assumption that the functional

$$w : \mathbb{R}^+ \to 2^{\mathbb{R}_0^+}, \quad \alpha \mapsto \left\{ \Psi(x_\alpha^\delta) \mid x_\alpha^\delta \in \mathcal{M}_{\alpha,y^\delta} \right\}$$

is single-valued and hence continuous (cf. Corollary 4.14) in [37, 40, 64]. For the subdifferential of $\tilde{m}(\alpha)$ to exist we need to show that $\tilde{m}(\alpha)$ is convex which is equivalent to $m(\alpha)$ being concave.

Lemma 4.19. *The functional $m(\alpha)$ defined in (4.5) is concave.*

Proof. Let $0 < \alpha < \beta$ and $0 < t < 1$, then for arbitrary minimizers $x_\alpha^\delta \in \mathcal{M}_{\alpha,y^\delta}$, $x_t \in \mathcal{M}_{t\alpha+(1-t)\beta,y^\delta}$ and $x_\beta^\delta \in \mathcal{M}_{\beta,y^\delta}$ we have

$$tm(\alpha) + (1-t)m(\beta)$$
$$= t\left(\left\| F(x_\alpha^\delta) - y^\delta \right\|^q + \alpha \Psi(x_\alpha^\delta) \right) + (1-t)\left(\left\| F(x_\beta^\delta) - y^\delta \right\|^q + \beta \Psi(x_\beta^\delta) \right)$$
$$\leq t\left(\left\| F(x_t) - y^\delta \right\|^q + \alpha \Psi(x_t) \right) + (1-t)\left(\left\| F(x_t) - y^\delta \right\|^q + \beta \Psi(x_t) \right)$$
$$= m(t\alpha + (1-t)\beta).$$

□

The following result allows to express the subdifferential of $\tilde{m}(\alpha)$ as the interval spanned by the elements in $-w(\alpha)$.

Lemma 4.20. *Let $\tilde{m}(\alpha) = -m(\alpha)$. Then the subdifferential of the convex functional $\tilde{m}$ at $\alpha > 0$ is given by*

$$\partial \tilde{m}(\alpha) = [-\sup w(\alpha), -\inf w(\alpha)].$$

Proof. We have seen in Lemma 4.19 that $m(\alpha)$ is concave and, consequently, $\tilde{m}(\alpha)$ is convex and hence the subdifferential exists. By Definition 1.21 of the subdifferential, we know that $\mu \in \partial \tilde{m}(\alpha)$ if and only if

$$\tilde{m}(\beta) - \tilde{m}(\alpha) \geq \mu\,(\beta - \alpha) \tag{4.16}$$

holds for all $\beta > 0$ and due to the definition of $\tilde{m}(\alpha)$ in (4.5) we can express the left hand side as

$$\tilde{m}(\beta) - \tilde{m}(\alpha) = m(\alpha) - m(\beta)$$
$$= \left\| F(x_\alpha^\delta) - y^\delta \right\|^q + \alpha \Psi(x_\alpha^\delta) - \left\| F(x_\beta^\delta) - y^\delta \right\|^q - \beta \Psi(x_\beta^\delta), \tag{4.17}$$

where $x_\alpha^\delta \in \mathcal{M}_{\alpha,y^\delta}$ and $x_\beta^\delta \in \mathcal{M}_{\beta,y^\delta}$ are arbitrary minimizers of the Tikhonov functionals corresponding to $\alpha, \beta > 0$. From their respective minimizing properties we obtain that

$$m(\alpha) \leq \left\| F(x_\beta^\delta) - y^\delta \right\|^q + \alpha \Psi(x_\beta^\delta)$$
$$m(\beta) \leq \left\| F(x_\alpha^\delta) - y^\delta \right\|^q + \beta \Psi(x_\alpha^\delta).$$

Using these inequalities together with (4.17) we derive

$$(\beta - \alpha)\left(-\Psi(x_\alpha^\delta)\right) \leq \tilde{m}(\beta) - \tilde{m}(\alpha) \leq (\beta - \alpha)\left(-\Psi(x_\beta^\delta)\right). \tag{4.18}$$

According Corollary 4.13 there exist minimizers x_1, x_2 in $\mathcal{M}_{\alpha,y^\delta}$ such that

$$\Psi(x_1) = \sup\ w(\alpha)$$
$$\Psi(x_2) = \inf\ w(\alpha).$$

In the light of the left hand side inequality in (4.18), therefore $\mu_1 = -\Psi(x_1)$ and $\mu_2 = -\Psi(x_2)$ belong to the subdifferential $\partial \tilde{m}(\alpha)$ and Proposition 1.22 asserts that the subdifferential is convex so that

$$I := [-\sup w(\alpha), -\inf w(\alpha)] \subset \partial \tilde{m}(\alpha).$$

It remains to show that no $\mu \notin I$ belongs to the subgradient $\partial \tilde{m}(\alpha)$. Note, that we can exclude positive μ from our considerations as according to Lemma 4.7 $m(\alpha)$ is monotonically increasing and hence $\tilde{m}(\alpha)$ is decreasing everywhere on $\mathbb{R}^+$. We distinguish the cases $\mu < -\sup\ w(\alpha)$ and $-\inf\ w(\alpha) < \mu \leq 0$.

In the first case in order to construct a value $\beta > 0$ such that (4.16) fails to hold we choose a strictly monotonically increasing, positive sequence $\alpha_n \to \alpha^-$. From Lemmata 4.7 and 4.12 we obtain $\sup w(\alpha_n) \geq \sup w(\alpha)$ for all $n \in \mathbb{N}$ and

$$\lim_{n\to\infty}\ \sup w(\alpha_n) = \sup w(\alpha) < -\mu.$$

This yields the existence of $N \in \mathbb{N}$ such that $\sup w(\alpha_N) < -\mu$ and if we set $\beta = \alpha_N < \alpha$ then it follows from the right hand side inequality in (4.18) that

$$\tilde{m}(\beta) - \tilde{m}(\alpha) \leq (\alpha - \beta) \ \Psi(x_\beta^\delta)$$
$$\leq (\alpha - \beta) \ \sup w(\beta)$$
$$< \mu \ (\beta - \alpha)$$

holds and thus $\mu \notin \partial \tilde{m}(\alpha)$.

Similarly, if $-\inf w(\alpha) < \mu$ then we choose any decreasing sequence $\alpha_n \to \alpha^+$ and observe that $\inf w(\alpha_n) \geq \inf w(\alpha)$ and

$$\lim_{n \to \infty} \inf w(\alpha_n) = \inf w(\alpha) > -\mu.$$

Setting, as before, $\beta = \alpha_N > \alpha$ for N large enough we again arrive at

$$\tilde{m}(\beta) - \tilde{m}(\alpha) \leq (\beta - \alpha) \left(-\Psi(x_\beta^\delta) \right)$$
$$\leq (\beta - \alpha) \left(-\inf w(\beta) \right)$$
$$< \mu \ (\beta - \alpha)$$

since $\beta - \alpha$ has switched signs and the proof is complete. □

Using the discrepancy functional

$$g(\alpha) = \left\{ \left\| F(x_\alpha^\delta) - y^\delta \right\| \ \middle| \ x_\alpha^\delta \in \mathcal{M}_{\alpha, y^\delta} \right\}$$

and the relation $g(\alpha) = (m(\alpha) - \alpha w(\alpha))^{1/q}$ in (4.6) we know from Definition 4.3 that α is chosen according to MDP if and only if

$$m(\alpha) - \alpha w(\alpha) \cap [(\tau_1 \delta)^q, (\tau_2 \delta)^q] \neq \emptyset.$$

Thus, Lemma 4.20 suggests that one might try to find such α as solutions of

$$m(\alpha) + \alpha \, \partial \tilde{m}(\alpha) \cap [(\tau_1 \delta)^q, (\tau_2 \delta)^q] \neq \emptyset. \qquad (4.19)$$

It can not be guaranteed that solutions α of (4.19) really admit a minimizer $x_\alpha^\delta \in \mathcal{M}_{\alpha, y^\delta}$ fulfilling

$$\tau_1 \delta \leq \left\| F(x_\alpha^\delta) - y^\delta \right\| \leq \tau_2 \delta.$$

In this sense (4.19) is a necessary condition for candidates of parameters satisfying MDP, but it is not sufficient. However, note that due to the monotonicity and continuity results for $m(\alpha)$ and $w(\alpha)$ in Lemmata 4.7, 4.11 and Corollary 4.13 and the convexity of the subgradient $\partial \tilde{m}(\alpha)$ it is clear, that (4.19) will have a solution, even if MDP cannot be satisfied.

Remark 4.21. These observations, in particular (4.19), have served as a motivation for the so-called *model function* approach. If $w(\alpha)$ is single-valued then according to Lemma 4.20 so is $\partial \tilde{m}(\alpha)$ and it follows that $m(\alpha)$ is differentiable. Conceptually, one now tries to find a 'simple' differentiable functional $M(\alpha)$ that locally models the behavior of $m(\alpha)$ – for example, the asymptotic properties in Lemma 4.15 – and to then (iteratively) find solutions α of

$$M(\alpha) + \alpha M'(\alpha) \cap [(\tau_1 \delta)^q, (\tau_2 \delta)^q] \neq \emptyset.$$

This algorithm has been introduced and improved for classical Tikhonov regularization where $\Psi(x) = \|x\|^2$ in [40, 64] and recently also suggested for more general penalty terms in [37]. ∎

Chapter 5

The Discrepancy Principle

In this chapter we complete our results from Chapter 3, where it has been shown that Tikhonov regularization in combination with Morozov's discrepancy principle (MDP) is a convergent regularization method. One question that was left open though is the existence of such an $\alpha = \alpha(\delta, y^\delta)$ fulfilling the discrepancy principle. On the grounds of the analytic properties of the discrepancy functionals g, w and m that we collected in the preceding chapter, we will now formulate a sufficient condition for MDP. Another issue that was not discussed in Chapter 3 is that according to Definition 2.1 of a convergent regularization method a suitable parameter choice rule should satisfy $\alpha(\delta, y^\delta) \to 0$ as $\delta \to 0$. Even though this was not necessary for the proof of convergence of Tikhonov regularization with parameter chosen according to MDP in Section 3.2, we show in this chapter that the asymptotic relations $\alpha(\delta, y^\delta) \to 0$ and $\delta^q/\alpha(\delta, y^\delta) \to 0$ hold true as $\delta \to 0$ for differentiable operators. In addition, we present some related formulations of MDP that are used in the literature and discuss a numeric algorithm for the computation of such a parameter $\alpha = \alpha(\delta, y^\delta)$ that relies on the monotonicity of the discrepancy functionals.

5.1 Existence of α according to MDP

Recall the definition of the discrepancy functionals in Chapter 4: For fixed $y \in \text{rg}(F)$, $\delta > 0$ and $y^\delta \in Y$ with $\|y - y^\delta\| \leq \delta$, we write

$$g(\alpha) = \left\{ G(x_\alpha^\delta) \mid x_\alpha^\delta \in \mathcal{M}_{\alpha, y^\delta} \right\}, \tag{5.1}$$

$$w(\alpha) = \left\{ \Psi(x_\alpha^\delta) \mid x_\alpha^\delta \in \mathcal{M}_{\alpha, y^\delta} \right\}, \tag{5.2}$$

$$m(\alpha) = J_{\alpha, y^\delta}(x_\alpha^\delta), \qquad x_\alpha^\delta \in \mathcal{M}_{\alpha, y^\delta}, \tag{5.3}$$

where $G(x)$ is defined as

$$G(x) = \left\| F(x) - y^\delta \right\|. \tag{5.4}$$

A regularization parameter $\alpha = \alpha(\delta, y^\delta)$ according to MDP has to fulfill

$$g(\alpha) \cap [\tau_1 \delta, \tau_2 \delta] \neq \emptyset$$

(compare to Definition 4.3). Bearing in mind the monotonicity of $g(\alpha)$ (cf. Lemma 4.7), such a parameter should thus exist, roughly speaking, if $g(\alpha)$ does not 'jump over' the interval $[\tau_1 \delta, \tau_2 \delta]$. For the precise formulation of this statement and its proof we will need the next Proposition from [1] which generalizes a result for classical Tikhonov regularization from [47].

Proposition 5.1. *Assume that*

$$\inf_{x \in X} G(x) < \tau_1 \delta < G(0), \qquad (5.5)$$

then there exist $\underline{\alpha}, \bar{\alpha} \in \mathbb{R}^+$ *such that*

$$\sup g(\underline{\alpha}) < \tau_1 \delta < \inf g(\bar{\alpha}).$$

Proof. From the asymptotic results in Lemma 4.15 and (5.5) we know that

$$\lim_{\alpha \to 0^+} g(\alpha) = \inf_{x \in X} G(x) < \tau_1 \delta,$$

where the limit applies to arbitrary cuts of $g(\alpha)$ in the sense of Definition 4.6. According to Lemma 4.12 $\sup g(\alpha) \in g(\alpha)$ for every $\alpha > 0$ and thus $\{\sup g(\alpha)\}_{\alpha > 0}$ is a cut of $\{g(\alpha)\}_{\alpha > 0}$. Consequently, there exists $\varepsilon > 0$ such that $\sup g(\alpha) < \tau_1 \delta$ for all $\alpha < \varepsilon$ and we may simply choose $\underline{\alpha} = \varepsilon/2$, for example.

To find $\bar{\alpha}$ we use again Lemma 4.15, namely,

$$\tau_1 \delta < G(0) = \lim_{\alpha \to \infty} g(\alpha).$$

Using the cut $\{\inf g(\alpha)\}_{\alpha > 0}$ of $\{g(\alpha)\}_{\alpha > 0}$ this readily ensures the existence of $\bar{\alpha}$ such that $\inf g(\bar{\alpha}) > \tau_1 \delta$. □

Remark 5.2. From the existence of a solution $x^\dagger$ of $F(x) = y$ we immediately obtain for noisy data y^δ with $\|y - y^\delta\| \leq \delta$ that the following chain of inequalities holds true:

$$\inf_{x \in X} G(x) = \inf_{x \in X} \left\| F(x) - y^\delta \right\| \leq \left\| F(x^\dagger) - y^\delta \right\| = \left\| y - y^\delta \right\| \leq \delta \leq \tau_1 \delta.$$

Therefore the assumption

$$\inf_{x \in X} G(x) < \tau_1 \delta$$

in Proposition 5.1 amounts to one of the above inequalities being strict, i.e., one of the following cases:

(i) There exists an element $\bar{x} \in X$ such that $F(\bar{x})$ approximates y^δ better than $y = F(x^\dagger)$ in the sense that

$$\left\|F(\bar{x}) - y^\delta\right\| < \left\|F(x^\dagger) - y^\delta\right\|.$$

(ii) The available noise bound δ is not sharp, i.e., $\left\|y - y^\delta\right\| < \delta$.

(iii) The constant τ_1 is strictly greater than 1.

We emphasize two sufficient conditions for the left hand side inequality in (5.5) that arise from this discussion: First of all, if y^δ belongs to the range of F and $\delta > 0$, then, clearly,

$$\inf_{x \in X} \left\|F(x) - y^\delta\right\| = 0 < \tau_1 \delta.$$

And secondly, if $y^\delta \notin \mathrm{rg}(F)$ and a best approximation of y^δ in $\mathrm{rg}(F)$ does not exist, then

$$\inf_{x \in X} \left\|F(x) - y^\delta\right\| < \left\|F(x^\dagger) - y^\delta\right\|.$$

However, if neither (i) nor (ii) can be ensured for the problem at hand, one may always choose the constant $\tau_1 > 1$. ∎

We are now ready to prove that the following condition is sufficient to ensure the existence of a regularization parameter α chosen according to the discrepancy principle in Definition 4.3.

Condition 5.3. Let $y \in \mathrm{rg}(F)$ and $\delta > 0$ be fixed. Assume that $y^\delta \in Y$ with $\left\|y - y^\delta\right\| \leq \delta$ satisfies

$$\inf_{x \in X} G(x) < \tau_1 \delta < G(0) \tag{5.6}$$

and that there is no $\alpha > 0$ such that

$$\inf g(\alpha) < \tau_1 \delta \leq \tau_2 \delta < \sup g(\alpha). \tag{5.7}$$

Note that (5.7) can be reformulated as: There is no $\alpha > 0$ with minimizers $x_1, x_2 \in \mathcal{M}_{\alpha, y^\delta}$ such that

$$\left\|F(x_1) - y^\delta\right\| < \tau_1 \delta \leq \tau_2 \delta < \left\|F(x_2) - y^\delta\right\|.$$

For the following theorem compare [47, Theorem 2.5] where the result is proven for the special case of classical Tikhonov regularization.

Theorem 5.4. *If Condition 5.3 is fulfilled, then there exists a parameter* $\alpha = \alpha(\delta, y^\delta) > 0$ *chosen according to Morozov's discrepancy principle, i.e.,*

$$g\left(\alpha(\delta, y^\delta)\right) \cap [\tau_1\delta, \tau_2\delta] \neq \emptyset \tag{5.8}$$

or, equivalently, there exists $x_\alpha^\delta \in \mathcal{M}_{\alpha(\delta,y^\delta), y^\delta}$ *such that*

$$\tau_1\delta \leq \left\|F(x_\alpha^\delta) - y^\delta\right\| \leq \tau_2\delta. \tag{5.9}$$

Proof. Assume that no α fulfilling (5.8) exists, and define

$$S = \{\alpha \in \mathbb{R}^+ \mid \sup g(\alpha) < \tau_1\delta\}$$
$$\tilde{S} = \{\alpha \in \mathbb{R}^+ \mid \inf g(\alpha) > \tau_1\delta\}.$$

Note that due to (5.6) Lemma 5.1 applies so that neither of these sets is empty. However, clearly, $S \cap \tilde{S} = \emptyset$ and we also show that

$$S \cup \tilde{S} = \mathbb{R}^+.$$

Indeed, $\alpha \in \mathbb{R}^+ \backslash (S \cup \tilde{S})$ if and only if

$$\inf g(\alpha) \leq \tau_1\delta \leq \sup g(\alpha),$$

but for such α either (5.8) holds – which is the case if $\inf g(\alpha) = \tau_1\delta$ or $\sup g(\alpha) \leq \tau_2\delta$ – or otherwise (5.7) in Condition 5.3 is violated.

Now, we define $\bar{\alpha} = \sup S$ and due to the monotonicity of $g(\alpha)$ (see Lemma 4.7) it follows that all $\alpha < \bar{\alpha}$ belong to S and all $\alpha > \bar{\alpha}$ to $\tilde{S}$. In addition, due to the above findings $\bar{\alpha}$ must belong to either S or $\tilde{S}$.

But using Lemma 4.12 it follows that in the first case we reach the contradiction

$$\tau_1\delta \leq \lim_{\alpha \to \bar{\alpha}^+} g(\alpha) = \sup g(\bar{\alpha}) < \tau_1\delta,$$

and, similarly, in the latter case

$$\tau_1\delta < \inf g(\bar{\alpha}) = \lim_{\alpha \to \bar{\alpha}^-} g(\alpha) \leq \tau_1\delta.$$

Therefore our initial assumption that there exists no α satisfying (5.8) must have been wrong. □

5.2 Asymptotics

We return once again to the question if Tikhonov regularization with parameter α chosen according to Morozov's discrepancy principle is a regularization method in the sense of Definition 2.1. What remains to show is that

$\alpha = \alpha(\delta, y^\delta) \to 0$ as $\delta \to 0$. As we have seen in Section 3.2 this property is not required in order to show (weak-*) convergence of $\mathcal{R}_\alpha(y^\delta) \to F^\dagger(y)$. Indeed, it is in general not necessarily true that $\alpha(\delta, y^\delta) \to 0$ as $\delta \to 0$, but as we will see the following condition is sufficient for that matter.

Condition 5.5. For all $x^\dagger \in \mathcal{L}$ (cf. Definition 2.13) we assume that

$$\liminf_{t \to 0^+} \frac{\left\| F((1-t)x^\dagger) - y \right\|^q}{t} = 0. \tag{5.10}$$

The following Lemma provides more insight as to the nature of Condition 5.5.

Lemma 5.6. *Let Y be a Hilbert space and $q > 1$. If $F(x)$ is differentiable in the directions $x^\dagger \in \mathcal{L}$ and the derivatives are uniformly bounded in a neigbourhood of $x^\dagger$, then Condition 5.5 is satisfied.*

Proof. It holds for any y (as long as it admits a Ψ-minimizing solution $x^\dagger$) that

$$\frac{d}{dt} \left\| F((1-t)x^\dagger) - y \right\|^q = q \left\| F((1-t)x^\dagger) - y \right\|^{q-2} \left\langle F'((1-t)x^\dagger)(-x^\dagger), F((1-t)x^\dagger) - y \right\rangle$$

and due to the boundedness of $F'((1-t)x^\dagger)$ near $t = 0$ this can be estimated by

$$\left| \frac{d}{dt} \left\| F((1-t)x^\dagger) - y \right\|^q \right| \leq q \left\| F((1-t)x^\dagger) - y \right\|^{q-1} \left\| F'((1-t)x^\dagger) \cdot x^\dagger \right\| \xrightarrow{t \to 0^+} 0,$$

since $\left\| F(x^\dagger) - y \right\| = 0$ by assumption. Together this yields

$$\lim_{t \to 0^+} \frac{\left\| F((1-t)x^\dagger) - y \right\|^q}{t} = \frac{d}{dt} \left\| F((1-t)x^\dagger) - y \right\|^q \bigg|_{t=0} = 0.$$

$\square$

Lemma 5.7. *Assume that Condition 5.5 is satisfied and that there exist $\alpha > 0$ and a solution x^* of $F(x) = y$ such that*

$$x^* = \arg\min_{x \in X} \left\{ \left\| F(x) - y \right\|^q + \alpha \Psi(x) \right\},$$

then $x^* = 0$.

Proof. Since $x^\star$ is a minimizer of $J_{\alpha,y}(x)$ with exact data y, we obtain for all $x^\dagger \in \mathcal{L}$
$$\alpha\Psi(x^\star) \leq \alpha\Psi(x^\dagger),$$
and this implies that $x^\star \in \mathcal{L}$. Due to the convexity of Ψ and to the fact that $0, x^\star \in \text{dom}(()\Psi)$ with $\Psi(0) = 0$, it holds for $t \in [0,1)$ that
$$\begin{aligned}\Psi((1-t)x^\star) &= \Psi((1-t)x^\star + t \cdot 0) \\ &\leq (1-t)\Psi(x^\star) + t\,\Psi(0) \\ &= (1-t)\Psi(x^\star).\end{aligned}$$
As $x^\star \in \mathcal{M}_{\alpha,y}$, we thus get
$$\begin{aligned}\alpha\Psi(x^\star) = J_{\alpha,y}(x^\star) &\leq J_{\alpha,y}((1-t)x^\star) \\ &\leq \|F((1-t)x^\star) - y\|^q + \alpha(1-t)\Psi(x^\star)\end{aligned}$$
and therefore
$$\alpha\,t\Psi(x^\star) \leq \|F((1-t)x^\star) - y\|^q.$$
Altogether this implies
$$0 \leq \alpha\Psi(x^\star) \leq \liminf_{t \to 0^+} \frac{\|F((1-t)x^\dagger) - y\|^q}{t} = 0,$$
which yields $\Psi(x^\star) = 0$. But according to Condtion 2.7 (i) this only holds if $x^\star = 0$. □

Remark 5.8. To illustrate that Lemma 5.7 does not hold for arbitrary F and y, we use the function
$$F(x) = 1 + \sqrt{|1-x|}, \quad x \in \mathbb{R},$$
which we will study in more detail in Section 5.1, as a simple counter example. Indeed, the derivative of F is unbounded in $x = 1$, so that Lemma 5.6 cannot be applied. For the choices $y = 1$, $q = 2$ and $\Psi(x) = |x|$ the unique solution of $F(x) = y$ is $x^\dagger = 1$ and it holds that
$$\lim_{t \to 0^+} \frac{|F((1-t)x^\dagger) - y|^2}{t} = \lim_{t \to 0^+} \frac{|t|}{t} = 1.$$
Therefore Condition 5.5 is violated and indeed for $\alpha = 1$ we find that
$$J_{1,y}(x) = (F(x) - y)^2 + \Psi(x) = |1-x| + |x| \geq 1 = J_{1,y}(x^\dagger) \quad \forall x \in \mathbb{R},$$
which shows that the Ψ-minimizing solution $x^\dagger \neq 0$ is also a minimizer of the Tikhonov functional for $\alpha > 0$. The same example also works in the classical Tikhonov case, choosing $\Psi(x) = x^2$. Note that for a choice $y > 1$,

Condition 5.5 is always satisfied, so that (5.10) truly depends on the exact data y. This is no longer an issue, however, if the Gâteux derivative of $F(x)$ is locally bounded (cf. Lemma 5.6). ∎

Theorem 5.9. *Let F, Ψ satisfy the Conditions 2.4, 2.7 and 5.5. Moreover, assume that data y^δ satisfy Condition 5.3 for all $\delta \in (0, \delta^*)$, where $\delta^* > 0$ is an arbitrary upper bound and that $\tau_1 > 1$ in Definition 3.3. Then any regularization parameters $\alpha = \alpha(\delta, y^\delta)$ chosen according to MDP satisfy*

$$\alpha(\delta, y^\delta) \to 0 \qquad \text{and} \qquad \frac{\delta^q}{\alpha(\delta, y^\delta)} \to 0 \qquad \text{as} \quad \delta \to 0.$$

Proof. Let $\delta^* > \delta_n \to 0$ and $\alpha_n = \alpha(\delta_n, y^{\delta_n})$ be chosen according to the discrepancy principle. As a shorthand we write $x_n = x_{\alpha_n}^{\delta_n}$ for the corresponding regularized solutions satisfying (3.2).

Assume that there is a subsequence of $\{\alpha_n\}$, denoted again by $\{\alpha_n\}$, and a constant $\bar{\alpha}$ such that $0 < \bar{\alpha} \leq \alpha_n \; \forall n$. If we denote the minimizers of $J_{\bar{\alpha}, y^{\delta_n}}(x)$ by

$$\bar{x}_n \in \mathcal{M}_{\bar{\alpha}, y^{\delta_n}} = \arg\min_{x \in X} \left\{ \left\| F(x) - y^{\delta_n} \right\|^q + \underline{\alpha} \Psi(x) \right\}$$

we obtain using Lemma 4.7 and (3.2) that

$$\left\| F(\bar{x}_n) - y^{\delta_n} \right\| \leq \left\| F(x_n) - y^{\delta_n} \right\| \leq \tau_2 \delta_n \to 0,$$

and

$$\limsup_{n \to \infty} \bar{\alpha} \Psi(\bar{x}_n) \leq \limsup_{n \to \infty} \left\{ \left\| F(\bar{x}_n) - y^{\delta_n} \right\| + \bar{\alpha} \Psi(\bar{x}_n) \right\}$$
$$\leq \bar{\alpha} \Psi(x^\dagger).$$

Therefore, $\{\bar{x}_n\}$ satisfies the assumptions of Lemma 3.5 and we can extract a subsequence $x_{n'} \rightharpoonup x^\dagger \in \mathcal{L}$. Because of the weak lower semi-continuity of Ψ and $\|F(\cdot) - y\|$, it holds that

$$\left\| F(x^\dagger) - y \right\|^q + \bar{\alpha} \Psi(x^\dagger) \leq \liminf_{n' \to \infty} \left(\left\| F(\bar{x}_{n'}) - y^{\delta_{n'}} \right\|^q + \bar{\alpha} \Psi(\bar{x}_{n'}) \right)$$
$$\leq \liminf_{n' \to \infty} \left(\left\| F(x) - y^{\delta_{n'}} \right\|^q + \bar{\alpha} \Psi(x) \right) \qquad \forall x \in X$$
$$= \| F(x) - y \|^q + \bar{\alpha} \Psi(x) \qquad \forall x \in X,$$

which shows that $x^\dagger$ is also a minimizer of $J_{\bar{\alpha}, y}(.)$. Therefore, $x^\dagger$ satisfies the assumptions of Lemma 5.7 and it follows that $x^\dagger = 0$, which in turn means that $y = F(0)$. This violates (5.6) in Condition 5.3, and we have reached a contradiction.

The second part of the theorem is an immediate consequence of (3.5) with $\tau_1 > 1$ and the assertion $\Psi(x_\alpha^\delta) \to \Psi(x^\dagger)$ in Theorem 3.6. ∎

Remark 5.10. In the proof of Theorem 5.9 we have used $\|F(0) - y\| > 0$, which is an immediate consequence of (5.6). On the other hand, whenever $\|F(0) - y\| > 0$ we can choose

$$0 < \delta^* \leq \frac{1}{\tau_2 + 1} \|F(0) - y\|$$

and for all $0 < \delta < \delta^*$ and y^δ satisfying $\|y - y^\delta\| \leq \delta$ we obtain

$$\left\|F(0) - y^\delta\right\| \geq \|F(0) - y\| - \left\|y - y^\delta\right\| \geq \|F(0) - y\| - \delta > \tau_2 \delta,$$

which is (5.6). Therefore (5.6) can be fulfilled for all δ smaller than some $\delta^* > 0$, whenever $y \neq F(0)$. ∎

5.3 Other formulations

Tikhonov et al. [62] provide a very rigorous analysis of variational methods for solving extremal problems which cover the case (3) under consideration in this survey. The authors discuss several different parameter choice rules, among them the following version of the discrepancy principle. Let $\alpha = \alpha(\delta, y^\delta)$ be chosen such that

$$\inf_{x_\alpha^\delta \in \mathcal{M}_{\alpha, y^\delta}} \left\|F(x_\alpha^\delta) - y^\delta\right\| \leq \delta \leq \sup_{x_\alpha^\delta \in \mathcal{M}_{\alpha, y^\delta}} \left\|F(x_\alpha^\delta) - y^\delta\right\|.$$

As opposed to our approach, the selection of the regularized solution x_α^δ corresponding to $\alpha = \alpha_a(\delta, y^\delta)$ is done depending on properties of arbitrarily chosen minimizers $x_{\alpha_1}^\delta, x_{\alpha_2}^\delta$ of (3) corresponding to $\alpha_1 = \alpha/r, \alpha_2 = r\alpha$ for some $r > 1$ and in a way such that the resulting discrepancy may be smaller than the noise level, i.e.

$$\left\|F(x_\alpha^\delta) - y^\delta\right\| \leq \delta$$

is possible, and no lower bound in terms of δ is available. Thus, in this case it can no longer be ensured that the discrepancy is of the same order as the noise level.

Another common formulation that has been considered, for example, by Engl et al. [17], is to choose

$$\alpha_b(\delta, y^\delta) = \sup\left\{\alpha > 0 \mid \forall x_\alpha^\delta \in \mathcal{M}_{\alpha, y^\delta} : \left\|F(x_\alpha^\delta) - y^\delta\right\| \leq \tau \delta\right\}.$$

This definition also carries the advantage that such a parameter α always exists and if we choose $\tau \in [\tau_1, \tau_2]$ from Definition 4.3, then it will also be admissible according to our definition of MDP provided that one of the corresponding minimizers $x_\alpha^\delta \in \mathcal{M}_{\alpha, y^\delta}$ satisfies (4.1). However, in itself it does not provide any information on how to choose x_α^δ among the set $\mathcal{M}_{\alpha,y^\delta}$, which is of great consequence to our analytic results.

The residual method consists in solving the constrained minimization problem

$$\Psi(x) \to \min \quad \text{subject to} \quad \left\| F(x) - y^\delta \right\| \leq \tau\delta$$

and it has been studied by Grasmair et al. in [25, 26]. For certain linear operators this method is equivalent to Tikhonov regularization if α is chosen according to MDP with $\tau_1 = \tau_2 = \tau$ (cf. [26, Proposition 2.2]). Using Lagrange multipliers it can be reformulated as the problem of minimizing

$$\tilde{J}_{\alpha,y^\delta}(x) = \left(\left\| F(x) - y^\delta \right\| - \tau\delta \right)^2 + \alpha\Psi(x),$$

that has been studied, e.g., also in [17, 46].

5.4 Computational aspects

One of the main challenges when using the discrepancy principle in practical application is the computation of a parameter α and a corresponding minimizer $x_\alpha^\delta \in \mathcal{M}_{\alpha,y^\delta}$ such that (4.1) is satisfied. For the solution of the optimization problem (2.5) many approaches can be found in the literature. We mention [7, 48]. It lies in the nature of a-posteriori parameter choice rules, that one has to compute a (sequence of) regularized solution(s), in our case minimizers of variational functionals of type (2.5), until the defining property of the parameter choice rule (4.1) is satisfied.

Here we will present one method to achieve this that was already suggested, e.g., in [61] and relies on the monotonicity of the discrepancy functionals in Lemma 4.7. It consists in choosing a starting value $\alpha_0 > 0$ and computing a corresponding minimizer x_0 of the Tikhonov functional. If the residual is too large,

$$\left\| F(x_0) - y^\delta \right\| > \tau_2\delta,$$

we know from Lemma 4.7 that all admissible values α according to the discrepandy principle are no larger than α_0 and we choose the next iterate according to

$$\alpha_1 = q\alpha_0$$

with some fixed constant $q < 1$. If, on the other hand, the residual is too small,

$$\left\| F(x_0) - y^\delta \right\| < \tau_1\delta,$$

we choose

$$\alpha_1 = \alpha_0/q.$$

Clearly, if
$$\tau_1 \delta \leq \left\| F(x_0) - y^\delta \right\| \leq \tau_2 \delta,$$
the initial values α_0, x_0 themselves satisfy MDP and our algorithm terminates.

We continue this procedure generating a sequence of iterates $\{(\alpha_k, x_k)\}$ until we either find a solution of (4.1) – in which case the algorithm terminates – or we arrive at a situation where
$$\left\| F(x_{k+1}) - y^\delta \right\| < \tau_1 \delta \leq \tau_2 \delta < \left\| F(x_k) - y^\delta \right\|.$$

Note that here the roles of α_k and α_{k+1} may be interchanged. At this point it is clear that
$$[\underline{\alpha}, \bar{\alpha}] \subset [\alpha_{k+1}, \alpha_k],$$
where $[\underline{\alpha}, \bar{\alpha}]$ denotes the set of parameters fulfilling the discrepancy principle. Thus, we continue with a bisection procedure choosing
$$\alpha_{k+2} = \frac{\alpha_k + \alpha_{k+1}}{2}.$$

Now, if the corresponding residual is between $\tau_1 \delta$ and $\tau_2 \delta$ the algorithm terminates. Otherwise, we use α_{k+2} as the new upper bound for $\bar{\alpha}$ or as a lower bound for $\underline{\alpha}$ if the residual is too big or small, respectively. The resulting algorithm is summarized in Algorithm 1.

The number of iterations – and hence the computational cost – of this algorithm depends on the quality of the initial guesses α_0, q for the problem at hand. In general, it will be cheaper to compute the minimizers of a Tikhonov functional with large parameter α, as the regularization term stabilizes the minimization process. This corresponds to a better conditioning in the linear case. However, if the starting values are such that $\alpha_0 \gg \bar{\alpha}$ and $q \approx 1$, then many iterations may be required to reach the interval of interest. On the other hand, if $\alpha_0 \approx \bar{\alpha}$ and $q \ll 1$ then possibly $\alpha_1 \ll \underline{\alpha}$ and we will need several very expensive iterations to get back. Therefore, an adaptive strategy might be advisable, which starts with $q \approx 1$ and consequently chooses q depending on the distance of the current residual to $\tau_1 \delta$ or $\tau_2 \delta$.

Another algorithm that has been proposed in the literature is the *model function* approach (see also Remark 4.21). It is based on the idea of approximating the discrepancy functional $m(\alpha)$ locally by a simpler function $M(\alpha)$ which maintains its basic properties and to then use this functional in order to find an approximate solution of the discrepancy condition (4.1). We refer the interested reader to [37, 40, 64] for a detailed description of the resulting algorithm and modifications thereof.

Algorithm 1 Monotone algorithm

Require: $\alpha_0 > 0$, $q < 1$

$k = 0$
$x_0 \in \arg\min_{x \in X} J_{\alpha_0, y^\delta}(x)$
$r_0 = \|F(x_0) - y^\delta\|$
while $r_k > \tau_2 \delta$ **do**
  $\alpha_{k+1} = q\alpha_k$
  $x_{k+1} \in \arg\min_{x \in X} J_{\alpha_{k+1}, y^\delta}(x)$
  $r_{k+1} = \|F(x_{k+1}) - y^\delta\|$
  $k = k + 1$
end while
if $k = 0$ **then**
  while $r_k < \tau_1 \delta$ **do**
    $\alpha_{k+1} = \alpha_k / q$
    $x_{k+1} \in \arg\min_{x \in X} J_{\alpha_{k+1}, y^\delta}(x)$
    $r_{k+1} = \|F(x_{k+1}) - y^\delta\|$
    $k = k + 1$
  end while
end if
while $r_k \notin [\tau_1 \delta, \tau_2 \delta]$ **do**
  $\alpha_{k+1} = (\alpha_k + \alpha_{k-1})/2$
  $x_{k+1} \in \arg\min_{x \in X} J_{\alpha_{k+1}, y^\delta}(x)$
  $r_{k+1} = \|F(x_{k+1}) - y^\delta\|$
  $k = k + 1$
end while
return (α_k, x_k)

5.5 An academic example

In this section we study Tikhonov regularization for a simple non-linear function. It will be shown that a regularization parameter α according to MDP exists for certain given data y^δ, whereas for others it may not. We explicitely compute the sets $g(\alpha)$ for different y^δ and $\alpha > 0$ and check if the defining property (4.7) of MDP is satisfied.

Let us consider the one dimensional non-linear function

$$F(x) = 1 + \sqrt{|1-x|}, \qquad x \in \mathbb{R}, \tag{5.11}$$

as an operator on the Hilbert space $\mathbb{R}$ and assume to be given only noisy data y^δ, where $\delta \geq |y - y^\delta|$ is known. We take as an approximate solution a minimizer x_α^δ of the ℓ_1-penalized Tikhonov functional

$$\begin{aligned} J_{\alpha,y^\delta}(x) &= \left|F(x) - y^\delta\right|^2 + \alpha\,|x| \\ &= (\sqrt{|1-x|} + (1-y^\delta))^2 + \alpha\,|x| \\ &= |1-x| + 2(1-y^\delta)\sqrt{|1-x|} + (1-y^\delta)^2 + \alpha\,|x|. \end{aligned}$$

This functional is differentiable everywhere except in $x \in \{0,1\}$. The derivative is given by

$$\frac{\partial}{\partial x} J_{\alpha,y^\delta}(x) = \operatorname{sign}(x-1)\left(1 + \frac{1-y^\delta}{\sqrt{|1-x|}}\right) + \alpha\operatorname{sign}(x). \tag{5.12}$$

and the first order necessary condition for a minimizer therefore reads

$$\left(1 + \alpha\operatorname{sign}(x_\alpha^\delta - 1)\operatorname{sign}(x_\alpha^\delta)\right)\sqrt{|1-x_\alpha^\delta|} = y^\delta - 1. \tag{5.13}$$

To find solutions to this equation we distinguish several cases and also consider the points of non-differentiability $x = 0$ and $x = 1$ separately.

Case 1. $x_\alpha^\delta < 0$. In this case (5.13) evaluates to

$$\sqrt{1-x_\alpha^\delta} = \frac{y^\delta - 1}{1+\alpha}$$

and for $y^\delta > 1$ we find the critical point

$$x_1 = 1 - \left(\frac{y^\delta - 1}{1+\alpha}\right)^2. \tag{5.14}$$

In order for this solution to also fulfill the assumption $x_\alpha^\delta < 0$ of this case, it is necessary that $\alpha < y^\delta - 2$. Given that we are only interested in values $\alpha > 0$ we need only consider x_1 whenever $y^\delta > 2$. The corresponding functional value is

$$J_{\alpha,y^\delta}(x_1) = (y^\delta - 1)^2 \frac{\alpha}{1+\alpha} - \alpha.$$

Case 2. $x_\alpha^\delta = 0$. The value of $J_{\alpha,y^\delta}(x)$ at $x_2 = 0$ is

$$J_{\alpha,y^\delta}(x_2) = 1 + 2(1 - y^\delta) + (1 - y^\delta)^2 = (2 - y^\delta)^2.$$

Case 3. $0 < x_\alpha^\delta < 1$. In this case (5.13) evaluates to

$$(1 - \alpha)\sqrt{1 - x_\alpha^\delta} = y^\delta - 1$$

and whenever $\text{sign}(y^\delta - 1) = \text{sign}(1 - \alpha) \neq 0$ we find the critical point

$$x_3 = 1 - \left(\frac{y^\delta - 1}{1 - \alpha}\right)^2. \tag{5.15}$$

In order for this solution to fulfill the defining assumption of this case, it has to hold that $|y^\delta - 1| < |1 - \alpha|$. Therefore we take x_3 into account in the following two cases:

- if $0 < y^\delta - 1 < 1 - \alpha$, i.e., $1 < y^\delta < 2$ and $0 < \alpha < 2 - y^\delta$.
- if $0 < 1 - y^\delta < \alpha - 1$, i.e., $y^\delta < 1$ and $2 - y^\delta < \alpha$.

The corresponding functional value is

$$J_{\alpha,y^\delta}(x_3) = (y^\delta - 1)^2 \frac{-\alpha}{1 - \alpha} + \alpha.$$

Case 4. $x_\alpha^\delta = 1$. The value of $J_{\alpha,y^\delta}(.)$ at $x_4 = 1$ is

$$J_{\alpha,y^\delta}(x_4) = (1 - y^\delta)^2 + \alpha.$$

Case 5. $x_\alpha^\delta > 1$. In this case (5.13) evaluates to

$$(1 + \alpha)\sqrt{x_\alpha^\delta - 1} = y^\delta - 1$$

and for $y^\delta > 1$ we find

$$x_5 = 1 + \left(\frac{y^\delta - 1}{1 + \alpha}\right)^2.$$

The corresponding functional value is

$$J_{\alpha,y^\delta}(x_5) = (y^\delta - 1)^2 \frac{\alpha}{1 + \alpha} + \alpha.$$

Note that

$$J_{\alpha,y^\delta}(x_2) = 1 + 2(1 - y^\delta) + (1 - y^\delta)^2 \leq (1 - y^\delta)^2 + \alpha = J_{\alpha,y^\delta}(x_4). \tag{5.16}$$

if and only if $\alpha \geq 3 - 2y^\delta$.

Lemma 5.11. *If $y^\delta > 2$, then the set of minimizers $\mathcal{M}_{\alpha,y^\delta}$ of $J_{\alpha,y^\delta}(x)$ is given by*

$$\mathcal{M}_{\alpha,y^\delta} = \begin{cases} \{x_1\} & 0 < \alpha < y^\delta - 2 \\ \{0\} & \alpha \geq y^\delta - 2, \end{cases}$$

with x_1 as defined in (5.14). If $1 < y^\delta < 2$, then

$$\mathcal{M}_{\alpha,y^\delta} = \begin{cases} \{x_3\} & 0 < \alpha < 2 - y^\delta \\ \{0\} & \alpha \geq 2 - y^\delta, \end{cases}$$

with x_3 as defined in (5.15). If $y^\delta = 1$, then

$$\mathcal{M}_{\alpha,y^\delta} = \begin{cases} \{1\} & 0 < \alpha < 1 \\ [0,1] & \alpha = 1 \\ \{0\} & \alpha > 1 \end{cases}$$

and if $y^\delta < 1$

$$\mathcal{M}_{\alpha,y^\delta} = \begin{cases} \{1\} & 0 < \alpha < 3 - 2y^\delta \\ \{0,1\} & \alpha = 3 - 2y^\delta \\ \{0\} & \alpha > 3 - 2y^\delta. \end{cases}$$

Proof. We compare those functional values of cases 1-5 above which are relevant to the α, y^δ at hand. In so doing for $y^\delta \geq 2$ and $0 < \alpha < y^\delta - 2$ we compare the values at $x = x_i, i \neq 3$. Clearly $J_{\alpha,y^\delta}(x_1) < J_{\alpha,y^\delta}(x_5) < J_{\alpha,y^\delta}(x_4)$ and moreover

$$J_{\alpha,y^\delta}(x_1) = (y^\delta - 1)^2 \frac{\alpha}{1+\alpha} - \alpha \leq (2 - y^\delta)^2 = J_{\alpha,y^\delta}(x_2).$$

holds since

$$0 < (1+\alpha)^2 + 2(1-y^\delta)(1+\alpha) + (1-y^\delta)^2 = (2 - y^\delta + \alpha)^2.$$

Therefore $x_1 = 1 - \left(\frac{y^\delta - 1}{1+\alpha}\right)^2$ is the global minimizer of $J_{\alpha,y^\delta}(.)$ in this case. For $\alpha \geq y^\delta - 2$ the necessary first order condition is no longer fulfilled at $x = x_1$ and comparing x_2 and x_5 we find that

$$J_{\alpha,y^\delta}(x_2) = (2 - y^\delta)^2 = 1 + 2(1 - y^\delta) + (1 - y^\delta)^2$$
$$< (y^\delta - 1)^2 \frac{\alpha}{1+\alpha} + \alpha = J_{\alpha,y^\delta}(x_5)$$

is equivalent to

$$0 < \alpha^2 - (y^\delta - 2)^2 + 2(y^\delta - 1)\alpha, \tag{5.17}$$

which holds true in this case, so that $x_2 = 0$ is the unique global minimizer for such α.

Now let $1 < y^\delta < 2$, then for $0 < \alpha < 2 - y^\delta$ candidates for a minimizer are the points $x_i, i > 1$. Here $J_{\alpha,y^\delta}(x_3) < J_{\alpha,y^\delta}(x_5) < J_{\alpha,y^\delta}(x_4)$ and

$$J_{\alpha,y^\delta}(x_3) = (y^\delta - 1)^2 \frac{-\alpha}{1-\alpha} + \alpha \leq (2 - y^\delta)^2 = J_{\alpha,y^\delta}(x_2)$$

holds because

$$0 < (1-\alpha)^2 + 2(1-y^\delta)(1-\alpha) + (1-y^\delta)^2 = (2 - \alpha - y^\delta)^2.$$

Therefore $x_3 = 1 - \left(\frac{y^\delta - 1}{1-\alpha}\right)^2$ is the unique minimizer of $J_{\alpha,y^\delta}(.)$ for $0 < \alpha < 2 - y^\delta$. If, on the other hand, $\alpha \geq 2 - y^\delta$, then x_3 drops out of the list of candidates and from (5.17) we see that $J_{\alpha,y^\delta}(x_2) < J_{\alpha,y^\delta}(x_5)$, so that again $x_2 = 0$ is the global minimizer.

The situation changes for $y^\delta = 1$, where

$$J_{\alpha,1}(x) = |1-x| + \alpha |x|$$

has the unique global minimizers $x_4 = 1$ if $\alpha < 1$ and $x_2 = 0$ if $\alpha > 1$, but is minimized by any value $x \in [0,1]$ if $\alpha = 1$.

In case that $y^\delta < 1$ and $\alpha < 2 - y^\delta$ we again only need to consider x_2 and x_4 and from (5.16) it follows that the unique minimizer is $x_4 = 1$. For $\alpha \geq 2 - y^\delta$ there is a critical point at $x = x_3$, but since in this situation clearly $J_{\alpha,y^\delta}(x_4) < J_{\alpha,y^\delta}(x_3)$, the unique minimizer remains at $x_4 = 1$ for $2 - y^\delta \leq \alpha < 3 - 2y^\delta$. For $\alpha = 3 - 2y^\delta$ the set of minimzers is $\mathcal{M}_{\alpha,y^\delta} = \{0,1\}$ and finally for $\alpha > 3 - 2y^\delta$ there is again a unique minimizer at $x_2 = 0$. □

Having computed the minimizers of $J_{\alpha,y^\delta}(x)$ we are now in position to look at the behaviour of the functional $G(x_\alpha^\delta)$ defined in (4.2) for different noisy data y^δ. We observe that for $y^\delta > 1$ there exists a unique global minimizer of $J_{\alpha,y^\delta}(x)$ for any value of α. Therefore the discrepancy functional

$$g : \alpha \mapsto \left\{ G(x_\alpha^\delta) : x_\alpha^\delta \in \mathcal{M}_{\alpha,y^\delta} \right\}$$

is single-valued everywhere and consequently according to Corollary 4.14 also continuous. In this case the noisy data belongs to the range of the operator, $y^\delta \in \mathrm{rg}(F) = [1,\infty)$, so that $g(\alpha) \to 0$ as $\alpha \to 0$. From Lemma 5.11 the exact values are found to be

$$g(\alpha) = \begin{cases} \frac{\alpha}{1+\alpha}(y^\delta - 1) & 0 < \alpha < y^\delta - 2 \\ y^\delta - 2 & \alpha \geq y^\delta - 2 \end{cases}$$

for $y^\delta > 2$, and

$$g(\alpha) = \begin{cases} \frac{\alpha}{1-\alpha}(y^\delta - 1) & 0 < \alpha < 2 - y^\delta \\ 2 - y^\delta & \alpha \geq 2 - y^\delta \end{cases}$$

for $1 < y^\delta < 2$. Note that in the limiting case $y^\delta = 2$ the functional $g(\alpha)$ vanishes for all $\alpha > 0$ and that this case indeed has to be excluded from our considerations as it violates (5.6) in Condition 5.3 due to $F(0) = y^\delta$. For any $y^\delta \in (1, \infty) \setminus \{2\}$ the discrepancy principle is applicable for δ sufficiently small. In Figure 5.1 we plot $g(\alpha)$ for data $y^\delta = 3$ and 1.1.

If $y^\delta = 1$, then $g(\alpha)$ is no longer single-valued at $\alpha = 1$. It is given by

$$g(\alpha) = \begin{cases} 0 & 0 < \alpha < 1 \\ [0,1] & \alpha = 1 \\ 1 & \alpha > 1. \end{cases}$$

Nevertheless g remains continuous in the sense that to each $t \in [0,1]$ there exists $x \in \mathcal{M}_{1,y^\delta} = [0,1]$ such that $\left|F(x) - y^\delta\right| = t$. Therefore the discrepancy principle remains applicable for all $0 < \delta < 1/\tau_2$. Figure 5.2 shows $g(\alpha)$ for data $y^\delta = 1$.

The situation changes completely though, when $y^\delta \notin \text{rg}(F)$, i.e., $y^\delta < 1$. The discrepancy functional is multi-valued at $\alpha = 3 - 2y^\delta$, where it consists of a discrete set of two isolated points rather than of an interval as in the previous case $y^\delta = 1$.

$$g(\alpha) = \begin{cases} 1 - y^\delta & 0 < \alpha < 3 - 2y^\delta \\ \{1 - y^\delta, 2 - y^\delta\} & \alpha = 3 - 2y^\delta \\ 2 - y^\delta & \alpha > 3 - 2y^\delta. \end{cases}$$

Since the exact data y is known to belong to $\text{rg}(F) = [1, \infty)$ the noise bound has to fulfill $\delta \geq 1 - y^\delta$. Now, if $\tau_1 \delta > 1$ and $\tau_2 \delta < 2 - y^\delta$, then the sufficient Condition 5.3 is violated and indeed there exists no parameter $\alpha > 0$ and minimizer $x^\delta_\alpha \in \mathcal{M}_{\alpha,y^\delta}$ satisfying the discrepancy principle. It therefore only remains applicable in the following cases

- $\delta = 1 - y^\delta$ and $\tau_1 = 1$, which readily implies that $y = 1$ as it is the only element in $\text{rg}(F)$ to comply with this noise bound. In this case we may choose any value $\alpha \in (0, 3 - 2y^\delta]$ and $x^\delta_\alpha = 1$ as the corresponding minimizer to find that the discrepancy principle is fulfilled and that, moreover, we have obtained the unique solution $x^\dagger = 1$ of $F(x) = y$.

- $\tau_2 \geq (2 - y^\delta)/\delta$. Here we may choose $\alpha \in [3 - 2y^\delta, \infty)$ and the minimizer $x^\delta_\alpha = 0$ so that the discrepancy principle is fulfilled. Note that if y^δ remains outside of $\text{rg}(F)$, then $\frac{2-y^\delta}{\delta} \geq 1/\delta \to \infty$ as $\delta \to 0$, so that eventually any constant τ_2 will be too small. However $y^\delta \notin \text{rg}(F)$ for all $\delta > 0$ is only possible if y belongs to the boundary of $\text{rg}(F)$, i.e., $y = 1$. Otherwise for δ sufficiently small the discrepancy principle will again be applicable once $y^\delta \geq 1$.

Figure 5.2 shows $g(\alpha)$ for data $y^\delta = 0.9$. We summarize our findings in the following Lemma.

Lemma 5.12. *Let F be as defined in (5.11). If $y^\delta \in [1,\infty)\backslash\{2\}$, then to any $\tau_2 \geq \tau_1 \geq 1$ and $0 < \delta < \left|2-y^\delta\right|/\tau_2$ there exists $\alpha = \alpha(\delta, y^\delta)$ and $x_\alpha^\delta \in \mathcal{M}_{\alpha,y^\delta}$ such that*

$$\tau_1 \delta \leq \left\|F\left(x_\alpha^\delta\right) - y^\delta\right\| \leq \tau_2 \delta$$

and therefore Morozov's discrepancy principle is applicable.

If $y^\delta < 1$, then the discrepancy principle is only applicable if $\tau_1 \delta = 1 - y^\delta$ or $\tau_2 \delta \geq 2 - y^\delta$.

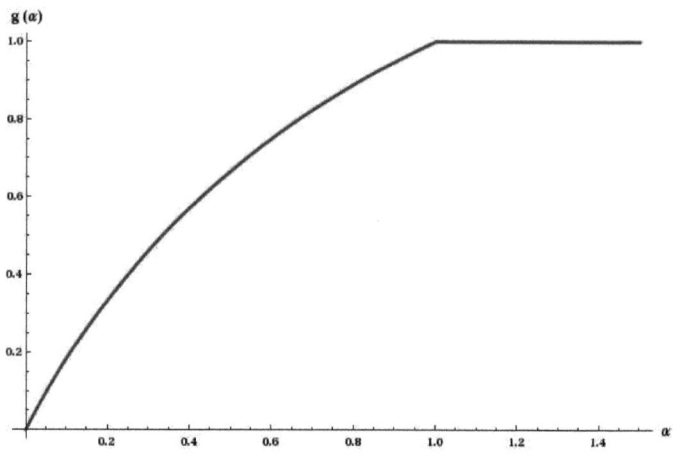

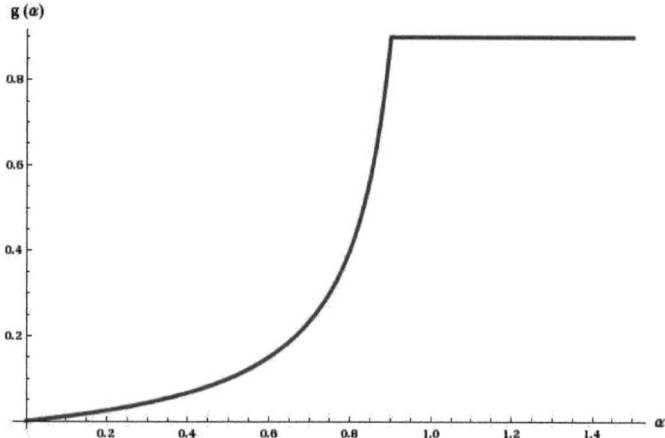

Figure 5.1: The function $g(\alpha)$ for $y^\delta = 3$ (top) and $y^\delta = 1.1$ (bottom)

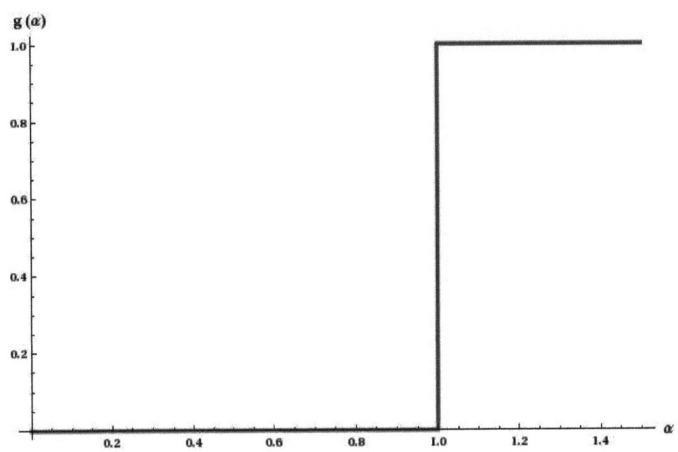
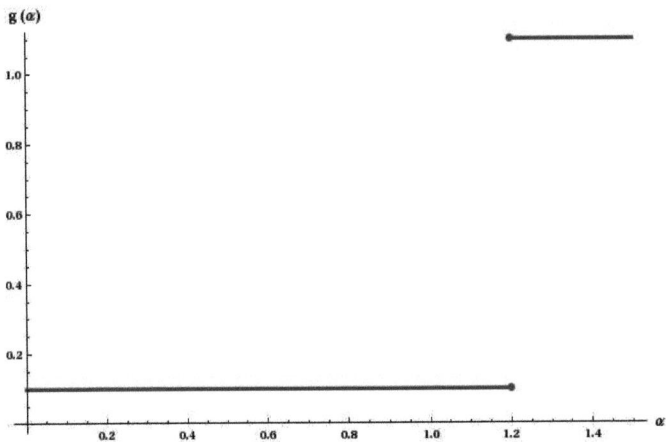

Figure 5.2: The function $g(\alpha)$ for $y^\delta = 1$ (top) and $y^\delta = 0.9$ (bottom)

Chapter 6

Convergence rates

We have seen that Tikhonov regularization with general, convex penalty term and regularization parameter chosen according to Morozov's discrepancy priniciple is a convergent regularization method. In this chapter we turn our attention to convergence rates. It is a well-known result due to Schock [57] that without any further knowledge of properties of the unknown solution $x^\dagger$ the speed of convergence may, in general, be arbitrarily slow. Additional assumptions on $x^\dagger$ that imply convergence rates are known as *source conditions*.

For Tikhonov regularization in Hilbert spaces convergence rates were first derived for solutions that are sufficiently smooth. One way of quantifying such smoothness properties is to assume that the solution belongs to some Sobolev space. If the operator under consideration is itself smoothing – which is certainly the case for integral operators, for example – then another way is to assume that the solution belongs to the range of the operator (which might, in turn, be a Sobolev space). In a Banach space setting with convex and differentiable penalty term $\Psi(x)$ and linear operator F this condition translates to

$$\Psi'(x^\dagger) \in \operatorname{rg}(F^*), \qquad (6.1)$$

which is known as a *range inclusion* type source condition. In Example 6.43 below we will discuss an even more general formulation of this condition for nonlinear operators and penalty terms that are no longer differentiable and we will show that (6.1) implies the existence of a constant $\beta > 0$ such that

$$-\langle \Psi'(x^\dagger), x - x^\dagger \rangle \leq \beta \left\| F(x) - F(x^\dagger) \right\|, \qquad \forall x \in X. \qquad (6.2)$$

For certain operators (6.1) and (6.2) are actually equivalent (cf. [33]). Inequalities of this type are known as *variational inequalities* and have been studied, e.g., in the recent works [5, 18, 24, 33, 35]. In what follows, we will illustrate how convergence rate results can be derived for source conditions expressed as variational inequalities.

6.1 Sublevelsets of $J_{\alpha,y}(x)$ and $\Psi(x)$

A main advantages of the formulation (6.2) over (6.1) is that even in Banach spaces it has several possible generalizations: One may consider powers of the norm on the right hand side of (6.2) and add other terms to it. Or, one may restrict the set of elements $x \in X$ for which (6.2) is required to hold. As we will see, choosing suitable sets may lead to considerable simplifications in the theoretic results.

6.1.1 Sublevelsets of $\Psi(x)$

We now introduce the domains on which the variational inequalities will be required to hold in Section 6.3. They consist of sublevelsets of the penalty term $\Psi(x)$, or more precisely, subsets thereof.

Definition 6.1. Let $\mathcal{L}$, $\psi^\dagger$ and $\mathcal{D}$ as defined in (2.3), (2.4) and (2.6), respectively. For $\rho \geq 0$ we denote

$$V_{\mathcal{L}}(\rho) = \left\{ x \in \mathcal{D} \mid \Psi(x) \leq \psi^\dagger \wedge \|F(x) - y\| \leq \rho \right\}. \tag{6.3}$$

It is easy to see that the sets $V_{\mathcal{L}}(\rho)$ are nested and contain $\mathcal{L}$ as the common intersection.

Lemma 6.2. *If $0 \leq \rho_1 \leq \rho_2$, then $V_{\mathcal{L}}(\rho_1) \subset V_{\mathcal{L}}(\rho_2)$ and $V_{\mathcal{L}}(0) = \mathcal{L}$.*

Proof. By definition of the Ψ-minimizing solutions, it holds for all $x^\dagger \in \mathcal{L}$ that

$$\Psi(x^\dagger) = \psi^\dagger, \qquad F(x^\dagger) = y$$

and that there is no solution $\bar{x}$ of $F(x) = y$ with $\Psi(x) < \psi^\dagger$. This readily establishes that $V_{\mathcal{L}}(0) = \mathcal{L}$.

Moreover, if $x \in V_{\mathcal{L}}(\rho_1)$, then clearly $x \in V_{\mathcal{L}}(\rho_2)$ which proves the monotonicity of the sets. □

It is an integral part of our analysis that the regularized solutions $x_\alpha^\delta \in \mathcal{M}_{\alpha,y^\delta}$ belong to the sets $V_{\mathcal{L}}(\rho)$ for any $\rho > 0$ once δ is small enough.

Lemma 6.3. *Let $0 < \delta \leq \rho/(\tau_2 + 1)$ and y^δ be such that $\|y - y^\delta\| \leq \delta$. If $\alpha = \alpha(\delta, y^\delta)$ is chosen according to MDP and $x_\alpha^\delta \in \mathcal{M}_{\alpha,y^\delta}$ fulfills (4.1), then*

$$x_\alpha^\delta \in V_{\mathcal{L}}(\rho).$$

Proof. According to Lemma 4.5 and Lemma 4.4 for α and $x_\alpha^\delta \in \mathcal{M}_{\alpha,y^\delta}$ fulfilling (4.1) we know

$$\Psi(x_\alpha^\delta) \leq \psi^\dagger$$

and

$$\left\|F(x_\alpha^\delta) - y\right\| \leq (\tau_2 + 1)\delta \leq \rho.$$

76

Here and below we denote by $B_\varepsilon(x^\dagger)$ the open ball with radius ε centered at $x^\dagger$. Under certain conditions the regularized solutions $x_\alpha^\delta \in \mathcal{M}_{\alpha,y^\delta}$ even belong to localized subsets of $V_\mathcal{L}(\rho)$.

Lemma 6.4. *Let the penalty term $\Psi(x)$ fulfill Condition 3.7 and the set $\mathcal{L}$ of Ψ-minimizing solutions consists of $x^\dagger$ only. Then to every $\varepsilon > 0$ there exists $\delta^* > 0$ such that for all $0 < \delta \leq \delta^*$ and y^δ with $\left\|y - y^\delta\right\| \leq \delta$ it holds that*
$$x_\alpha^\delta \in B_\varepsilon(x^\dagger) \cap V_\mathcal{L}(\rho).$$
Here, $\alpha = \alpha(\delta, y^\delta)$ is chosen according to MDP and $x_\alpha^\delta \in \mathcal{M}_{\alpha,y^\delta}$ satisfies (4.1).

Proof. Theorem 3.8 asserts that $x_\alpha^\delta \to x^\dagger$ under the assumptions of the Lemma and since $x_\alpha^\delta \in V_\mathcal{L}(\rho)$ according to Lemma 6.3 the existence of such δ^* follows. □

6.1.2 Sublevelsets of $J_{\alpha,y}(x)$

In what follows we will assume variational inequalities to hold on a set $V_\mathcal{L}(\rho)$ as in (6.3) for some $\rho > 0$. Alternatively, one could work on sublevelsets of the Tikhonov functional with exact data

$$S_\alpha(\sigma) = \left\{x \in \mathcal{D} \mid J_{\alpha,y}(x) \leq \sigma\right\}, \tag{6.4}$$

which have been used, for example, in [5]. We will show in Section 6.1.3 that these sublevelsets contain a set $V_\mathcal{L}(\rho)$ for all values $\sigma > \alpha\psi^\dagger$ and ρ small enough. But before doing so, we establish how the $S_\alpha(\sigma)$ are nested with respect to α and σ and that they contain the sets $\mathcal{M}_{\alpha,y^\delta}$ if $\sigma > \alpha\psi^\dagger$.

Lemma 6.5. *If $\alpha > 0$ is fixed and $0 \leq \sigma_1 \leq \sigma_2$, then $S_\alpha(\sigma_1) \subset S_\alpha(\sigma_2)$. However, if $\sigma > 0$ is fixed and $0 \leq \alpha_1 \leq \alpha_2$, then $S_{\alpha_2}(\sigma) \subset S_{\alpha_1}(\sigma)$.*

Proof. Assume α is fixed and let $x \in S_\alpha(\sigma_1)$, then clearly
$$J_{\alpha,y}(x) \leq \sigma_1 \leq \sigma_2,$$
so that $x \in S_\alpha(\sigma_2)$. If, on the other hand, σ is fixed and $x \in S_{\alpha_2}(\sigma)$, then
$$J_{\alpha_1,y}(x) \leq J_{\alpha_2,y}(x) \leq \sigma,$$
which shows that such x also belongs to $S_{\alpha_1}(\sigma)$. □

The following Lemma asserts that for a choice $\sigma < \alpha\psi^\dagger$ the sublevelsets $S_\alpha(\sigma)$ do not contain any of the true solutions $x^\dagger \in \mathcal{L}$ and such σ would therefore be too small for the purpose of obtaining convergence rates.

Lemma 6.6. *If $\sigma < \alpha \psi^\dagger$, then $\mathcal{L} \cap S_\alpha(\sigma) = \emptyset$.*

Proof. Let $x^\dagger \in \mathcal{L}$ be an arbitrary Ψ-minimizing solution. Then

$$J_{\alpha,y}(x^\dagger) = \left\|F(x^\dagger) - y\right\|^q + \alpha \Psi(x^\dagger) = \alpha \psi^\dagger > \sigma$$

and thus $x^\dagger \notin S_\alpha(\sigma)$. $\square$

In [5] it was proven that the sublevelsets $S_{\bar{\alpha}}(\sigma)$ contain the regularized solutions $x_\alpha^\delta \in \mathcal{M}_{\alpha,y^\delta}$ for δ small enough, if σ is chosen as

$$\sigma = \bar{\alpha}(1 + \psi^\dagger)$$

and $\alpha = \alpha(\delta)$ or $\alpha = \alpha(\delta, y^\delta)$ is any parameter choice rule fulfilling $\alpha \to 0$ and $\delta^q/\alpha \to 0$ as $\delta \to 0$, which is the case for MDP (cf. Theorem 5.9). We repeat the proof here and show that actually to any $\sigma > \bar{\alpha}\psi^\dagger$ there exists $\bar{\delta} > 0$ such that the regularized solutions $x_\alpha^\delta \in \mathcal{M}_{\alpha,y^\delta}$ belong to the sets $S_{\bar{\alpha}}(\sigma)$ whenever $0 < \delta \leq \bar{\delta}$.

Lemma 6.7. *Let $\bar{\alpha} > 0$ and $\sigma > \bar{\alpha}\psi^\dagger$ be arbitrary but fixed and let $\alpha = \alpha(\delta, y^\delta)$ be such that*

$$\alpha \to 0 \quad \text{and} \quad \frac{\delta^q}{\alpha} \to 0 \quad \text{as} \quad \delta \to 0,$$

where $\|y - y^\delta\| \leq \delta$. Then there exists $\bar{\delta} > 0$ such that $\mathcal{M}_{\alpha,y^\delta} \subset S_{\bar{\alpha}}(\sigma)$ for all $0 < \delta < \bar{\delta}$.

Proof. Let $c = \max(2^{q-1}, 1)$, then it holds for any $x_\alpha^\delta \in \mathcal{M}_{\alpha,y^\delta}$ that

$$\begin{aligned}
J_{\bar{\alpha},y}(x_\alpha^\delta) &= \left\|F(x_\alpha^\delta) - y\right\|^q + \bar{\alpha}\Psi(x_\alpha^\delta) \\
&\leq \left(\left\|F(x_\alpha^\delta) - y^\delta\right\| + \delta\right)^q + \bar{\alpha}\Psi(x_\alpha^\delta) \\
&\leq c\left\|F(x_\alpha^\delta) - y^\delta\right\|^q + c\delta^q + \bar{\alpha}\Psi(x_\alpha^\delta) \\
&\leq cJ_{\alpha,y^\delta}(x_\alpha^\delta) + c\delta^q + (\bar{\alpha} - c\alpha)\Psi(x_\alpha^\delta).
\end{aligned}$$

Thus, if we choose $\delta_1 > 0$ small enough that $c\alpha < \bar{\alpha}$ for all $\delta < \delta_1$, then due to $J_{\alpha,y^\delta}(x_\alpha^\delta) \leq J_{\alpha,y^\delta}(x^\dagger)$ and $\Psi(x_\alpha^\delta) \leq J_{\alpha,y^\delta}(x^\dagger)/\alpha$ we obtain

$$\begin{aligned}
J_{\bar{\alpha},y}(x_\alpha^\delta) &\leq cJ_{\alpha,y^\delta}(x_\alpha^\delta) + c\delta^q + (\bar{\alpha} - c\alpha)\Psi(x_\alpha^\delta) \\
&\leq c(\delta^q + \alpha\psi^\dagger) + c\delta^q + (\bar{\alpha} - c\alpha)\left(\frac{\delta^q}{\alpha} + \psi^\dagger\right) \\
&\leq c\delta^q + \bar{\alpha}\frac{\delta^q}{\alpha} + \bar{\alpha}\psi^\dagger.
\end{aligned}$$

Knowing that $\delta^q/\alpha \to 0$ as $\delta \to 0$ and $\sigma > \bar{\alpha}\psi^\dagger$ we find $\delta_2 > 0$ small enough such that for all $\delta < \delta_2$

$$0 < c\delta^q + \bar{\alpha}\frac{\delta^q}{\alpha} \leq \sigma - \bar{\alpha}\psi^\dagger$$

so that $J_{\bar{\alpha},y}(x_\alpha^\delta) \leq \sigma$ whenever $0 < \delta < \bar{\delta} := \min(\delta_1, \delta_2)$, which is to say that $x_\alpha^\delta \in S_{\bar{\alpha}}(\sigma)$. □

6.1.3 Relation between the sublevelsets of $\Psi(x)$ and $J_{\alpha,y}(x)$

Let us now establish a relation between the sets $V_{\mathcal{L}}(\rho)$ and $S_\alpha(\sigma)$ for different values of ρ, α and σ.

Lemma 6.8. *If $\alpha > 0$ is fixed and $\sigma \geq \alpha\psi^\dagger$, then*

$$V_{\mathcal{L}}(\rho) \subset S_\alpha(\sigma),$$

whenever $\rho \leq (\sigma - \alpha\psi^\dagger)^{1/q}$. Conversely,

$$S_\alpha(\sigma) \subset V_{\mathcal{L}}(\rho),$$

whenever $\psi^\dagger > 0$, $\alpha \geq \rho^q/\psi^\dagger > 0$ and $\sigma \leq \rho^q$.

Proof. If $\sigma \geq \alpha\psi^\dagger$ and $x \in V_{\mathcal{L}}(\rho)$ for $\rho \leq (\sigma - \alpha\psi^\dagger)^{1/q}$, then

$$\begin{aligned} J_{\alpha,y}(x) &= \|F(x) - y\|^q + \alpha\Psi(x) \\ &\leq \rho^q + \alpha\psi^\dagger \\ &\leq \sigma, \end{aligned}$$

so that $x \in S_\alpha(\sigma)$. On the other hand, if $x \in S_\alpha(\sigma)$, then

$$\|F(x) - y\|^q + \alpha\Psi(x) \leq \sigma,$$

yields the relations

$$\|F(x) - y\|^q \leq \sigma \quad \text{and} \quad \alpha\Psi(x) \leq \sigma$$

whence it follows for $\alpha \geq \rho^q/\psi^\dagger > 0$ and $\sigma \leq \rho^q$ that

$$\|F(x) - y\| \leq \sigma^{1/q} \leq \rho,$$

and

$$\rho^q \frac{\Psi(x)}{\psi^\dagger} \leq \alpha\Psi(x) \leq \sigma \leq \rho^q.$$

Since $\rho^q > 0$ according to our assumptions, we obtain $\Psi(x) \leq \psi^\dagger$ and, consequently, $x \in V_{\mathcal{L}}(\rho)$. □

Note, that if y^δ satisfies (5.6), then $x = 0$ cannot be a solution of $F(x) = y$ and due to Condition 2.7 (i) it follows that $\psi^\dagger > 0$. In other words, if $x = 0$ were a solution of $F(x) = y$ it would necessarily be the unique Ψ-minimizing solution and $\psi^\dagger = \Psi(0) = 0$. This situation corresponds to the case that there is more noise than signal in the data, which we may exclude from our considerations by means of (5.6), so that $\psi^\dagger > 0$ is guaranteed for $x^\dagger \in \mathcal{L}$.

Lemma 6.8 shows that for different choices of ρ, α and σ the sets $V_\mathcal{L}(\rho)$ are contained in $S_\alpha(\sigma)$ or the other way around. However, we have only seen that $S_\alpha(\sigma) \subset V_\mathcal{L}(\rho)$ if $\sigma \leq \rho^q \leq \alpha\psi^\dagger$ and if one of these inequalities is strict, then Lemma 6.6 states that such $S_\alpha(\sigma)$ does not contain any true solutions $x^\dagger \in \mathcal{L}$. Moreover, the sets $V_\mathcal{L}(\rho)$ do in general not contain any sublevelset $S_\alpha(\sigma)$ for a combination of parameters α and σ such that $\sigma > \alpha\psi^\dagger$. This can be seen from the following counterexample. It is in this sense that the sets $V_\mathcal{L}(\rho)$ are smaller than the sets $S_\alpha(\sigma)$.

Lemma 6.9. *Let F be a linear operator and let the penalty term $\Psi(x)$ be p-homogeneous (with $p > 0$), i.e.,*

$$\Psi(tx) = t^p \Psi(x), \qquad \forall t \in \mathbb{R}_0^+, x \in X.$$

If $\alpha, \psi^\dagger > 0$ and $\sigma > \alpha\psi^\dagger$, then there exists $\bar{c} > 1$ such that for all $x^\dagger \in \mathcal{L}$ it holds that $\bar{c}x^\dagger \in S_\alpha(\rho)$, but there is no $\rho \geq 0$ such that $\bar{c}x^\dagger \in V_\mathcal{L}(\rho)$.

Proof. According to our assumption $\sigma > \alpha\psi^\dagger > 0$, it is

$$c = \frac{\sigma}{\alpha\psi^\dagger} > 1$$

and we may choose $d > 1$ such that $d^p \in (1, c)$. Consequently, in case $y \neq 0$

$$\varepsilon = \frac{(\sigma - d^p\alpha\psi^\dagger)^{1/q}}{\|y\|} > \frac{(\sigma - c\alpha\psi^\dagger)^{1/q}}{\|y\|} = 0,$$

so that $\bar{c} = \min(d, 1 + \varepsilon) > 1$. Due to our choices it follows for arbitrary $x^\dagger \in \mathcal{L}$ that

$$\begin{aligned} J_{\alpha,y}(\bar{c}x^\dagger) &= \left\|F(\bar{c}x^\dagger) - y\right\|^q + \alpha\Psi(\bar{c}x^\dagger) \\ &= \left\|\bar{c}F(x^\dagger) - y\right\|^q + \bar{c}^p\alpha\Psi(x^\dagger) \\ &= (\bar{c} - 1)^q \|y\|^q + \bar{c}^p\alpha\psi^\dagger \\ &\leq \varepsilon^q \|y\|^q + d^p\alpha\psi^\dagger \\ &\leq \sigma, \end{aligned}$$

and $\bar{c}x^\dagger \in S_\alpha(\sigma)$. Note, that in case $y = 0$ we may simple choose $\bar{c} = d$ to obtain

$$J_{\alpha,y}(\bar{c}x^\dagger) = \bar{c}^p\alpha\psi^\dagger \leq \sigma.$$

However, in both of the above cases $\bar{c} > 1$ and thus

$$\Psi(\bar{c}x^\dagger) = \bar{c}^p \psi^\dagger > \psi^\dagger$$

which shows that $\bar{c}x^\dagger \notin V_{\mathcal{L}}(\rho)$ for any $\rho > 0$. □

6.2 Bregman and Taylor distance

In order to formulate the variational inequalities as well as to measure and estimate convergence rates, we will make use of the following distances. Our first quantitative estimates on the distance between the regularized solutions and a Ψ-minimizing solution $x^\dagger \in \mathcal{L}$ will be given with respect to the generalized Bregman distance, which is defined as follows.

Definition 6.10. Let $\partial \Psi(x) \subset X^*$ denote the subgradient of Ψ at $x \in X$. The generalized Bregman distance with respect to Ψ of two elements $x, z \in X$ is defined as

$$D_\Psi(x, z) = \{D_\Psi^\xi(x, z) : \xi \in \partial \Psi(z) \neq \emptyset\},$$

where

$$D_\Psi^\xi(x, z) = \Psi(x) - \Psi(z) - \langle \xi, x - z \rangle \tag{6.5}$$

denotes the Bregman distance with respect to Ψ and $\xi \in \partial \Psi(z)$. We remark that throughout this section $\langle \cdot, \cdot \rangle$ denotes the dual pairing in X^*, X or Y^*, Y and not the inner product on a Hilbert space. Moreover $\|\cdot\|_{Y^*}$ denotes the norm on Y^* and, in accordance with our previous notations, we write $\|\cdot\|$ for the norms in the Banach spaces X and Y.

Throughout the remainder of this chapter we will assume that the operator $F : X \to Y$ is Fréchet differentiable at arbitrary but fixed $x^\dagger \in \mathcal{L}$. We start by introducing the following notational conventions.

Definition 6.11. For $x \in X, x^\dagger \in \mathcal{L}$, we denote the norm of the second order Taylor remainder by

$$\mathcal{T}(x, x^\dagger) = \left\| F(x) - F(x^\dagger) - F'(x^\dagger)(x - x^\dagger) \right\|, \tag{6.6}$$

and call

$$D_\mathcal{T}\left(x, x^\dagger\right) = \left\| F'(x^\dagger)(x - x^\dagger) \right\| \tag{6.7}$$

the *Taylor distance* of x and $x^\dagger$.

6.3 Variational inequalities as source and nonlinearity conditions on sublevelsets

In the recent work [5] Boţ and Hofmann formulated the conjecture, that convergence rate results cannot be proven if the operator F fails to satisfy a structural condition of the form

$$D_{\mathcal{T}}\left(x, x^\dagger\right) \leq C\varphi\left(\left\|F(x) - F(x^\dagger)\right\|\right), \quad (6.8)$$

for $x \in S_\alpha(\sigma)$ as defined in (6.4), where $\sigma = \alpha(1 + \psi^\dagger)$, φ is a continuous, strictly increasing function through the origin and $C > 0$. We now introduce variational inequalities which are generalizations of the standard source and nonlinearity conditions discussed in Example 6.43 (i) and (ii) below, and which ultimately also fall into the framework of (6.8), but restricted to the smaller sets $V_\mathcal{L}(\rho)$ defined in (6.3) or subsets thereof. Similar inequalities were also used in the recent works [5, 18, 24, 35].

Condition 6.12. (VIE) Assume that for $x^\dagger \in \mathcal{L}$ there exist $\xi \in \partial\Psi(x^\dagger)$, $0 < \kappa \leq 1$, $\rho > 0$ and $\beta_i, \gamma_i \geq 0, i = 1, 2, 3$, such that

$$-\left\langle \xi, x - x^\dagger \right\rangle \leq \beta_1 D_\Psi^\xi\left(x, x^\dagger\right) + \beta_2 D_\mathcal{T}\left(x, x^\dagger\right) + \beta_3 \left\|F(x) - F(x^\dagger)\right\|^\kappa \quad (6.9)$$

$$\mathcal{T}\left(x, x^\dagger\right) \leq \gamma_1 D_\Psi^\xi\left(x, x^\dagger\right) + \gamma_2 D_\mathcal{T}\left(x, x^\dagger\right) + \gamma_3 \left\|F(x) - F(x^\dagger)\right\|^\kappa, \quad (6.10)$$

holds for all $x \in V_\mathcal{L}(\rho)$ and the constants β_i, γ_i fulfill

$$\beta_1 < 1, \quad \gamma_2 < 1 \quad \text{and} \quad \frac{\beta_2 \gamma_1}{(1 - \beta_1)(1 - \gamma_2)} < 1. \quad (6.11)$$

Applications where variational inequalities as in Condition 6.12 are fulfilled are, for example, phase retrieval problems and inverse option pricing, which were studied in [33].

Remark 6.13. If the variational inequalities (6.9) and (6.10) hold on sublevelsets $S_{\bar{\alpha}}(\sigma)$ as defined in (6.4), with

$$\sigma > \bar{\alpha}\psi^\dagger,$$

then according to Lemma 6.8 there exists $\rho > 0$ such that $V_\mathcal{L}(\rho) \subset S_{\bar{\alpha}}(\sigma)$ and the inequalities thus especially hold for all $x \in V_\mathcal{L}(\rho)$. ∎

When using parameter choice rules other than the discrepancy principle, e.g. a-priori rules, then Lemma 4.5 may no longer hold true and the only information concerning the relation between the values of the penalty term at the regularized solutions x_α^δ and the Ψ-minimizing solutions $x^\dagger$ we might have at hand is Lemma 3.5, namely that

$$\lim_{\delta \to 0} \Psi(x_\alpha^\delta) = \Psi(x^\dagger).$$

This is, however, not sufficient to ensure that $x_\alpha^\delta \in V_\mathcal{L}(\rho)$ as defined in (6.3). Therefore when working with other parameter choice rules one often times needs to consider larger sets

$$\tilde{V}_\mathcal{L}(\rho, \eta) = \left\{ x \in \mathcal{D} \mid \Psi(x) \leq \psi^\dagger + \eta \text{ and } \left\| F(x) - F(x^\dagger) \right\| \leq \rho \right\} \quad (6.12)$$

for some $\eta > 0$ (see, for example, [24]). Even though this distinction is seemingly small, most of the terms in Condition 6.12 become redundant when working with MDP.

Lemma 6.14. *Condition 6.12 is equivalent to: Assume that for $x^\dagger \in \mathcal{L}$ there exist $\xi \in \partial\Psi(x^\dagger)$, $0 < \kappa \leq 1$, $\rho > 0$ and $\tilde{\beta}_3, \tilde{\gamma}_3 \geq 0$ such that*

$$-\langle \xi, x - x^\dagger \rangle \leq \tilde{\beta}_3 \left\| F(x) - F(x^\dagger) \right\|^\kappa \quad (6.13)$$

$$\mathcal{T}(x, x^\dagger) \leq \tilde{\gamma}_3 \left\| F(x) - F(x^\dagger) \right\|^\kappa \quad (6.14)$$

holds for all $x \in V_\mathcal{L}(\rho)$.

Proof. Note first, that if (6.13) and (6.14) hold for $\tilde{\beta}_3, \tilde{\gamma}_3 \geq 0$, then Condition 6.12 is clearly fulfilled with $\beta_1 = \beta_2 = \gamma_1 = \gamma_2 = 0$, $\beta_3 = \tilde{\beta}_3$ and $\gamma_3 = \tilde{\gamma}_3$.

To show the other implication, we write $\beta_i' = \beta_i/(1-\beta_1)$ and $\gamma_i' = \gamma_i/(1-\gamma_2)$ for $i = 1, 2, 3$. For $x \in V_\mathcal{L}(\rho)$ we have using $\Psi(x) \leq \psi^\dagger$

$$D_\Psi^\xi \left(x, x^\dagger \right) = \Psi(x) - \Psi(x^\dagger) - \langle \xi, x - x^\dagger \rangle$$

$$\leq -\langle \xi, x - x^\dagger \rangle.$$

Note that for $x \in V_\mathcal{L}(\rho)$ and $0 < \kappa \leq 1$

$$\left\| F(x) - F(x^\dagger) \right\| \leq \max(1, \rho) \left\| F(x) - F(x^\dagger) \right\|^\kappa$$

holds, because if $\left\| F(x) - F(x^\dagger) \right\| \leq 1$, then

$$\left\| F(x) - F(x^\dagger) \right\| \leq \left\| F(x) - F(x^\dagger) \right\|^\kappa$$

and if $\rho > 1$ and $1 \leq \left\| F(x) - F(x^\dagger) \right\| \leq \rho$, then

$$\left\| F(x) - F(x^\dagger) \right\| \leq \rho \leq \rho \left\| F(x) - F(x^\dagger) \right\|^\kappa.$$

Thus, using (6.9), $\beta_1 < 1$ and
$$D_{\mathcal{T}}\left(x, x^\dagger\right) \leq \mathcal{T}(x, x^\dagger) + \left\|F(x) - F(x^\dagger)\right\|$$
we obtain for $x \in V_{\mathcal{L}}(\rho)$
$$\begin{aligned}
-\langle \xi, \; x - x^\dagger \rangle &\leq \beta_2' D_{\mathcal{T}}\left(x, x^\dagger\right) + \beta_3' \left\|F(x) - F(x^\dagger)\right\|^\kappa \\
&\leq \beta_2' \left(\mathcal{T}(x, x^\dagger) + \left\|F(x) - F(x^\dagger)\right\|\right) \\
&\quad + \beta_3' \left\|F(x) - F(x^\dagger)\right\|^\kappa \\
&\leq \beta_2' \mathcal{T}(x, x^\dagger) + (\beta_2' \max(1, \rho) + \beta_3') \left\|F(x) - F(x^\dagger)\right\|^\kappa.
\end{aligned}$$
Consequently, if (6.10) holds and $\gamma_2 < 1$, one gets
$$\begin{aligned}
\mathcal{T}(x, x^\dagger) &\leq -\gamma_1' \langle \xi, \; x - x^\dagger \rangle + \gamma_2' \left\|F(x) - F(x^\dagger)\right\| \\
&\quad + \gamma_3' \left\|F(x) - F(x^\dagger)\right\|^\kappa \\
&\leq \gamma_1' \beta_2' \mathcal{T}(x, x^\dagger) \\
&\quad + \left((\gamma_1' \beta_2' + \gamma_2') \max(1, \rho) + \gamma_1' \beta_3' + \gamma_3'\right) \left\|F(x) - F(x^\dagger)\right\|^\kappa
\end{aligned}$$
and due to (6.11) we find that (6.13) and (6.14) hold with
$$\tilde{\gamma}_3 = \frac{(\gamma_1' \beta_2' + \gamma_2') \max(1, \rho) + \gamma_1' \beta_3' + \gamma_3'}{1 - \gamma_1' \beta_2'}$$
and
$$\tilde{\beta}_3 = \beta_2' \tilde{\gamma}_3 + \beta_2' \max(1, \rho) + \beta_3'.$$
$\square$

Lemma 6.15. *If Condition 6.12 holds, there exist $\bar{\beta}_3, \bar{\gamma}_3 \geq 0$, such that for all $x \in V_{\mathcal{L}}(\rho)$*
$$D_\Psi^\xi\left(x, x^\dagger\right) \leq \bar{\beta}_3 \left\|F(x) - F(x^\dagger)\right\|^\kappa \tag{6.15}$$
$$D_{\mathcal{T}}\left(x, x^\dagger\right) \leq \bar{\gamma}_3 \left\|F(x) - F(x^\dagger)\right\|^\kappa. \tag{6.16}$$
hold.

Proof. We choose $\tilde{\beta}_3, \tilde{\gamma}_3$ as in Lemma 6.14 and obtain that for all $x \in V_{\mathcal{L}}(\rho)$
$$D_\Psi^\xi\left(x, x^\dagger\right) \leq -\langle \xi, x - x^\dagger \rangle \leq \tilde{\beta}_3 \left\|F(x) - F(x^\dagger)\right\|^\kappa \tag{6.17}$$

As in the proof of Lemma 6.14 we use that for $x \in V_{\mathcal{L}}(\rho)$ and $0 < \kappa \leq 1$

$$\left\|F(x) - F(x^\dagger)\right\| \leq \max(1,\rho) \left\|F(x) - F(x^\dagger)\right\|^\kappa,$$

whence in combination with (6.14) it follows that

$$D_{\mathcal{T}}\left(x, x^\dagger\right) \leq \mathcal{T}(x, x^\dagger) + \left\|F(x) - F(x^\dagger)\right\|$$
$$\leq (\tilde{\gamma}_3 + \max(1,\rho)) \left\|F(x) - F(x^\dagger)\right\|^\kappa.$$

Consequently, setting $\bar{\beta}_3 = \tilde{\beta}_3$ and $\bar{\gamma}_3 = \tilde{\gamma}_3 + \max(1,\rho)$ finishes the proof. □

6.4 Convergence rates in Bregman and Taylor distance

In [1] it has been proven that for the parameter choice rule MDP, the source condition from Example 6.43 (i) and a nonlinearity condition as in (6.46) yield a convergence rate of order $\mathcal{O}(\delta)$ in the Bregman distance. We will now show that similar results still hold under the more general Condition 6.12 with respect to the Bregman distance and also in the Taylor distance $D_{\mathcal{T}}\left(x, x^\dagger\right)$.

Theorem 6.16. *Let Condition 6.12 hold for* $x^\dagger \in \mathcal{L}, \xi \in \partial \Psi(x^\dagger)$. *If* $\alpha = \alpha(\delta, y^\delta)$ *is chosen according to MDP then for* $x_\alpha^\delta \in \mathcal{M}_{\alpha,y^\delta}$ *satisfying* (4.1), *it holds that*

$$D_\Psi^\xi\left(x_\alpha^\delta, x^\dagger\right) = \mathcal{O}(\delta^\kappa) \quad as \quad \delta \to 0, \tag{6.18}$$

$$D_{\mathcal{T}}\left(x_\alpha^\delta, x^\dagger\right) = \mathcal{O}(\delta^\kappa) \quad as \quad \delta \to 0. \tag{6.19}$$

Proof. According to Lemma 6.3 we know that $x_\alpha^\delta \in V_{\mathcal{L}}(\rho)$ whenever δ is small enough. Thus, if Condition 6.12 holds, we apply Lemma 6.15 to obtain $\bar{\beta}_3$ and $\bar{\gamma}_3$ such that

$$D_\Psi^\xi\left(x_\alpha^\delta, x^\dagger\right) \leq \bar{\beta}_3 \left\|F(x_\alpha^\delta) - F(x^\dagger)\right\|^\kappa$$
$$\leq \bar{\beta}_3 \tau_2^\kappa \delta^\kappa$$
$$= \mathcal{O}(\delta^\kappa) \quad as \quad \delta \to 0,$$

where the last estimate stems from the definition of MDP in (4.1). Similarly,

$$D_{\mathcal{T}}\left(x_\alpha^\delta, x^\dagger\right) \leq \bar{\gamma}_3 \left\|F(x_\alpha^\delta) - F(x^\dagger)\right\|^\kappa$$
$$\leq \bar{\gamma}_3 \tau_2^\kappa \delta^\kappa$$
$$= \mathcal{O}(\delta^\kappa) \quad as \quad \delta \to 0.$$

Remark 6.17. In the special case where X is a Hilbert space and $\Psi(x) = \|x\|^2$ it holds that $D_\Psi^\xi \left(x_\alpha^\delta, x^\dagger \right) = \left\| x_\alpha^\delta - x^\dagger \right\|^2$. Thus the convergence rate $\mathcal{O}(\delta)$ with respect to the Bregman distance in Theorem 6.16 corresponds to a rate $\mathcal{O}(\delta^{\kappa/2})$ in norm. ∎

Remark 6.18. We have shown in Lemma 6.15 that instead of assuming (6.9) and (6.10) to hold for elements $x \in V_\mathcal{L}(\rho)$ we might as well assume (6.15) and (6.16), which are a consequence of the first two inequalities. However, the later allow for the following generalization.

Assume that for $x^\dagger \in \mathcal{L}$ and $\xi \in \partial\Psi(x^\dagger)$ there exist $\rho > 0$ and $\beta, \gamma \geq 0$, such that for all $x \in V_\mathcal{L}(\rho)$

$$D_\Psi^\xi \left(x, x^\dagger \right) \leq \beta \, \varphi \left(\left\| F(x) - F(x^\dagger) \right\| \right)$$
$$D_\mathcal{T} \left(x, x^\dagger \right) \leq \gamma \, \varphi \left(\left\| F(x) - F(x^\dagger) \right\| \right).$$

holds, where $\varphi : [0, \infty) \to [0, \infty)$ is a continuous, strictly increasing function satisfying $\varphi(0) = 0$.

Condition 6.12 implies that such variational inequalities on the distances hold with $\varphi(t) = t^\kappa$ (cf. Lemma 6.15).

Under these generalized assumptions we would obtain the convergence rates

$$D_\Psi^\xi \left(x_\alpha^\delta, x^\dagger \right) = \mathcal{O}(\varphi(\delta)) \quad \text{as} \quad \delta \to 0,$$
$$D_\mathcal{T} \left(x_\alpha^\delta, x^\dagger \right) = \mathcal{O}(\varphi(\delta)) \quad \text{as} \quad \delta \to 0$$

arguing in the same fashion as in the proof of Theorem 6.16. ∎

6.5 Convergence rates in norm and the penalty term topology

To prove convergence rates in norm we introduce another variational inequality.

Condition 6.19. Assume that to $x^\dagger \in \mathcal{L}$ there exist $\xi \in \partial\Psi(x^\dagger), \mu_i \geq 0, r, \rho > 0$, and $0 < \kappa \leq 1$ such that

$$\Psi(x - x^\dagger)^r \leq \mu_1 D_\Psi^\xi \left(x, x^\dagger \right) + \mu_2 D_\mathcal{T} \left(x, x^\dagger \right) + \mu_3 \left\| F(x) - F(x^\dagger) \right\|^\kappa \tag{6.20a}$$

or
$$\left\|x - x^\dagger\right\|^r \leq \mu_1 D_\Psi^\xi\left(x, x^\dagger\right) + \mu_2 D_T\left(x, x^\dagger\right) + \mu_3 \left\|F(x) - F(x^\dagger)\right\|^\kappa \tag{6.20b}$$

holds for all $x \in V_\mathcal{L}(\rho)$.

In Chapter 7 below, we will see that in the context of sparse recovery Condition 6.19 is satisfied.

In combination with the variational source- and nonlinearity variational inequalities in Condition 6.12, we see from Lemma 6.14 that also Condition 6.19 simplifies considerably.

Lemma 6.20. *Let Conditions 6.12 and 6.19 hold for $x^\dagger \in \mathcal{L}$, then there exists a constant $\bar{\mu} \geq 0$ such that for all $x \in V_\mathcal{L}(\rho)$*

$$\Psi(x - x^\dagger)^r \leq \bar{\mu} \left\|F(x) - F(x^\dagger)\right\|^\kappa \tag{6.21a}$$

holds for (6.20a), or

$$\left\|x - x^\dagger\right\|^r \leq \bar{\mu} \left\|F(x) - F(x^\dagger)\right\|^\kappa \tag{6.21b}$$

for (6.20b).

Proof. Choose $\bar{\beta}$ and $\bar{\gamma}$ as in Lemma 6.14, then it follows from (6.20a) that

$$\Psi(x - x^\dagger)^r \leq \mu_1 D_\Psi^\xi\left(x, x^\dagger\right) + \mu_2 D_T\left(x, x^\dagger\right) + \mu_3 \left\|F(x) - F(x^\dagger)\right\|^\kappa$$
$$\leq \bar{\mu} \left\|F(x) - F(x^\dagger)\right\|^\kappa,$$

where
$$\bar{\mu} = \mu_1 \bar{\beta} + \mu_2 \bar{\gamma} + \mu_3.$$

Similarly, from (6.20b) we obtain (6.21b) for the same $\bar{\mu}$. □

When using MDP as the parameter choice rule either one of the additional variational inequalities (6.20a) or (6.20b) immediately yields convergence rates in some topology.

Theorem 6.21. *If Conditions 6.12 and 6.19 are fulfilled for $x^\dagger \in \mathcal{L}$ and $\alpha = \alpha(\delta, y^\delta)$ is chosen according to MDP, then with (6.20a) we have that*

$$\Psi(x_\alpha^\delta - x^\dagger) = \mathcal{O}(\delta^{\kappa/r}) \quad as \quad \delta \to 0 \tag{6.22}$$

and with (6.20b) that

$$\left\|x_\alpha^\delta - x^\dagger\right\|_X = \mathcal{O}(\delta^{\kappa/r}) \quad as \quad \delta \to 0. \tag{6.23}$$

holds for any $x_\alpha^\delta \in \mathcal{M}_{\alpha, y^\delta}$ satisfying (4.1).

Proof. The assumptions of Theorem 6.16 hold for $x = x_\alpha^\delta$ and from (6.20a) or (6.20b), respectively, we obtain using (6.18), (6.19) and (4.1) that

$$\Psi(x_\alpha^\delta - x^\dagger)^r \leq \mu_1 D_\Psi^\xi \left(x_\alpha^\delta, x^\dagger\right) + \mu_2 D_T \left(x_\alpha^\delta, x^\dagger\right) + \mu_3 \left\|F(x_\alpha^\delta) - F(x^\dagger)\right\|^\kappa$$
$$= \mathcal{O}(\delta^\kappa),$$

or

$$\left\|x_\alpha^\delta - x^\dagger\right\|^r \leq \mu_1 D_\Psi^\xi \left(x_\alpha^\delta, x^\dagger\right) + \mu_2 D_T \left(x_\alpha^\delta, x^\dagger\right) + \mu_3 \left\|F(x_\alpha^\delta) - F(x^\dagger)\right\|^\kappa$$
$$= \mathcal{O}(\delta^\kappa),$$

which are the desired convergence rates. □

Remark 6.22. Condtion 6.19 is satsified, for example, if $\Psi(x)$ is r-coercive, i.e., for some $c_r, \rho > 0$,

$$\left\|x - x^\dagger\right\|^r \leq c_r D_\Psi(x, x^\dagger) \qquad (6.24)$$

holds for all $x \in X$. It is well known, that the sparsity constraints $\Psi_{p,w}(x)$ with $1 \leq p \leq 2$ in Definition 2.17 fulfill (6.24) with $r = 2$, and for the optimal case $\kappa = 1$ we would obtain the classical rate

$$\left\|x_\alpha^\delta - x^\dagger\right\| = \mathcal{O}(\delta^{1/2}) \quad \text{as} \quad \delta \to 0. \qquad (6.25)$$

But – as we will see in Chapter 7 – even (6.20a) with $r = 1$ holds true in this setting and the resulting convergence result with respect to $\Psi_{p,w}$ is stronger than convergence in norm (cf. Lemma 2.20). Nevertheless, for different choices of the penalty term Ψ it may be more suitable to use (6.20b) instead. ∎

In some situations it is possible to simplify the assumptions and still obtain the same convergence rates. For example, if either $\mu_1 = 0$ or $\mu_2 = 0$ in Condition 6.19, then we don't need estimates on $D_\Psi^\xi \left(x, x^\dagger\right)$ or $D_T \left(x, x^\dagger\right)$, respectively, and we may dispense with (6.9) or (6.10) provided that these inequalities are not coupled.

Theorem 6.23. *Let Condition 6.19 be satisfied with $\mu_1 = 0$ and let (6.10) in Condition 6.12 hold with $\gamma_1 = 0$, then (6.22) follows with (6.20a) or (6.23) with (6.20b) for α chosen according to MDP and any $x_\alpha^\delta \in \mathcal{M}_{\alpha, y^\delta}$ satisfying (4.1).*

Proof. We use

$$\left\|F(x) - F(x^\dagger)\right\| \leq \max(1, \rho) \left\|F(x) - F(x^\dagger)\right\|^\kappa$$

from the proof of Lemma 6.14 to obtain with (6.10) that

$$D_{\mathcal{T}}\left(x,x^{\dagger}\right) \leq \mathcal{T}(x,x^{\dagger}) + \left\|F(x) - F(x^{\dagger})\right\|$$
$$\leq \gamma_2 D_{\mathcal{T}}\left(x,x^{\dagger}\right) + (\gamma_3 + \max(1,\rho)) \left\|F(x) - F(x^{\dagger})\right\|^{\kappa}$$

holds for all $x \in V_{\mathcal{L}}(\rho)$. Now, according to Lemma 6.3 $x_\alpha^\delta \in V_{\mathcal{L}}(\rho)$ and together with (6.20a) and (4.8) it follows that

$$\Psi(x_\alpha^\delta - x^\dagger)^r \leq \mu_2 D_{\mathcal{T}}\left(x_\alpha^\delta, x^\dagger\right) + \mu_3 \left\|F(x_\alpha^\delta) - F(x^\dagger)\right\|^\kappa$$
$$\leq \left(\mu_2 \frac{\gamma_3 + \max(1,\rho)}{1 - \gamma_2} + \mu_3\right)(\tau_2 \delta)^\kappa = \mathcal{O}(\delta^\kappa)$$

or, similarly, with (6.20b) that

$$\left\|x_\alpha^\delta - x^\dagger\right\|^r \leq \left(\mu_2 \frac{\gamma_3 + \max(1,\rho)}{1 - \gamma_2} + \mu_3\right)(\tau_2 \delta)^\kappa = \mathcal{O}(\delta^\kappa),$$

which yield the convergence rates. □

6.6 Local variational inequalities

The sets $V_{\mathcal{L}}(\rho)$ are chosen in a way that ensures $x_\alpha^\delta \in V_{\mathcal{L}}(\rho)$ if α satisfies MDP and δ is small enough. But Lemma 6.4 asserts that $x_\alpha^\delta \in B_\varepsilon(x^\dagger) \cap V_{\mathcal{L}}(\rho)$ under certain conditions and if this is the case it would suffice for (6.9) and (6.10) to hold on these sets only, where $\varepsilon > 0$ may be arbitrarily small. We briefly describe such a localized version of our results in this section.

Condition 6.24. (Local VIE) Assume that to $x^\dagger \in \mathcal{L}$ there exist $\xi \in \partial\Psi(x^\dagger)$, $0 < \kappa \leq 1$, $\varepsilon, \rho > 0$ and $\beta_i, \gamma_i \geq 0, i = 1,2,3$ fulfilling (6.11), such that (6.9) and (6.10) hold for $x \in B_\varepsilon(x^\dagger) \cap V_{\mathcal{L}}(\rho)$.

As in Lemma 6.14 we obtain the equivalence of Condition 6.24 to the following special case.

Condition 6.25. Assume that for $x^\dagger \in \mathcal{L}$ there exist $\xi \in \partial\Psi(x^\dagger)$, $0 < \kappa \leq 1$, $\rho > 0$ and $\tilde{\beta}_3, \tilde{\gamma}_3 \geq 0$ such that

$$-\langle \xi, x - x^\dagger \rangle \leq \tilde{\beta}_3 \left\|F(x) - F(x^\dagger)\right\|^\kappa$$
$$\mathcal{T}(x, x^\dagger) \leq \tilde{\gamma}_3 \left\|F(x) - F(x^\dagger)\right\|^\kappa$$

holds for all $x \in B_\varepsilon(x^\dagger) \cap V_{\mathcal{L}}(\rho)$.

The proof is carried out exactly as the proof of Lemma 6.14 only restricted to the subset $B_\varepsilon(x^\dagger) \cap V_\mathcal{L}(\rho)$. The same goes for the following Lemma.

Lemma 6.26. *If Condition 6.24 holds, there exist $\bar{\beta}_3, \bar{\gamma}_3 \geq 0$, such that for all $x \in B_\varepsilon(x^\dagger) \cap V_\mathcal{L}(\rho)$ it holds that*

$$D_\Psi^\xi\left(x, x^\dagger\right) \leq \bar{\beta}_3 \left\|F(x) - F(x^\dagger)\right\|^\kappa$$
$$D_\mathcal{T}\left(x, x^\dagger\right) \leq \bar{\gamma}_3 \left\|F(x) - F(x^\dagger)\right\|^\kappa.$$

Therefore we obtain the same convergence rate result in Bregman and Taylor distance for local variational inequalities as in the general case under the following additional assumptions.

Condition 6.27. *Let $\mathcal{L}$ in (2.3) consist of a unique element $x^\dagger$ and Ψ fulfill Condition 3.7.*

Theorem 6.28. *Let Condition 6.27 and 6.24 hold for $\xi \in \partial\Psi(x^\dagger)$. If $\alpha = \alpha(\delta, y^\delta)$ is chosen according to MDP then for $x_\alpha^\delta \in \mathcal{M}_{\alpha, y^\delta}$ satisfying (4.1), it holds that*

$$D_\Psi^\xi\left(x_\alpha^\delta, x^\dagger\right) = \mathcal{O}(\delta^\kappa) \quad \text{as} \quad \delta \to 0, \tag{6.26}$$
$$D_\mathcal{T}\left(x_\alpha^\delta, x^\dagger\right) = \mathcal{O}(\delta^\kappa) \quad \text{as} \quad \delta \to 0. \tag{6.27}$$

Proof. According to Lemma 6.4 we know that $x_\alpha^\delta \in B_\varepsilon(x^\dagger) \cap V_\mathcal{L}(\rho)$ whenever δ is small enough. Thus, if Condition 6.24 holds, we apply Lemma 6.26 to obtain $\bar{\beta}_3$ and $\bar{\gamma}_3$ such that

$$D_\Psi^\xi\left(x_\alpha^\delta, x^\dagger\right) \leq \bar{\beta}_3 \left\|F(x_\alpha^\delta) - F(x^\dagger)\right\|^\kappa \leq \bar{\beta}_3 \tau_2^\kappa \delta^\kappa = \mathcal{O}(\delta^\kappa)$$

as $\delta \to 0$, where the last estimate stems from the definition of MDP in (4.1). Similarly,

$$D_\mathcal{T}\left(x_\alpha^\delta, x^\dagger\right) \leq \bar{\gamma}_3 \left\|F(x_\alpha^\delta) - F(x^\dagger)\right\|^\kappa \leq \bar{\gamma}_3 \tau_2^\kappa \delta^\kappa = \mathcal{O}(\delta^\kappa)$$

as $\delta \to 0$ using again (4.1). □

We will use these convergence rates in combination with another variational inequality to obtain convergence rates in norm or the penalty term topology as we have done for the general case in Section 6.5.

Condition 6.29. *Assume that to $x^\dagger \in \mathcal{L}$ there exist $\xi \in \partial\Psi(x^\dagger)$, $\mu_i \geq 0$, $r, \varepsilon, \rho > 0$, and $0 < \kappa \leq 1$ such that either*

$$\Psi(x - x^\dagger)^r \leq \mu_1 D_\Psi^\xi\left(x, x^\dagger\right) + \mu_2 D_\mathcal{T}\left(x, x^\dagger\right) + \mu_3 \left\|F(x) - F(x^\dagger)\right\|^\kappa \tag{6.28a}$$

or

$$\left\|x - x^\dagger\right\|^r \le \mu_1 D_\Psi^\xi\left(x, x^\dagger\right) + \mu_2 D_\mathcal{T}\left(x, x^\dagger\right) + \mu_3 \left\|F(x) - F(x^\dagger)\right\|^\kappa \tag{6.28b}$$

holds for all $x \in B_\varepsilon(x^\dagger) \cap V_\mathcal{L}(\rho)$.

Condition 6.30. Assume that either one of the following statements holds.

(i) Let Ψ fulfill Condition 3.7, $\mathcal{L}$ in (2.3) consist of a unique element $x^\dagger$ and Condition 6.24 hold for $x^\dagger$.

(ii) Let Condition 6.12 hold for $x^\dagger \in \mathcal{L}$ and $\xi \in \partial\Psi(x^\dagger)$ such that the set

$$\mathcal{P} = \left\{z \in X \mid D_\Psi^\xi(z, x^\dagger) = D_\mathcal{T}(z, x^\dagger) = 0\right\} \tag{6.29}$$

consists of $x^\dagger$ only.

Note that $\mathcal{P} = \{x^\dagger\}$ is clearly satisfied for all $x^\dagger \in \mathcal{L}$ if Ψ is strictly convex or also for any $x^\dagger$ such that $F'(x^\dagger)$ is injective.

Theorem 6.31. *Let Conditions 6.30 and 6.29 hold. If $\alpha = \alpha(\delta, y^\delta)$ is chosen according to MDP then for $x_\alpha^\delta \in \mathcal{M}_{\alpha, y^\delta}$ satisfying (4.1)*

$$\Psi(x_\alpha^\delta - x^\dagger) = \mathcal{O}(\delta^{\kappa/r}) \quad \text{or} \quad \left\|x_\alpha^\delta - x^\dagger\right\| = \mathcal{O}(\delta^{\kappa/r}),$$

as $\delta \to 0$ hold for (6.28a) or (6.28b), respectively.

Proof. Condition 6.30 is made up of two cases, either of which implies that $x_\alpha^\delta \to x^\dagger$. Indeed, if Ψ fulfills Condition 3.7 and $\mathcal{L} = \{x^\dagger\}$, then convergence is established in Theorem 3.8. On the other hand, in the second case Theorem 6.16 asserts that $D_\Psi^\xi\left(x_\alpha^\delta, x^\dagger\right) \to 0$ and $D_\mathcal{T}\left(x_\alpha^\delta, x^\dagger\right) \to 0$ as $\delta \to 0$ and if $\mathcal{P} = \{x^\dagger\}$ then $x_\alpha^\delta \to x^\dagger$.

Thus, the estimates on $\Psi(x-x^\dagger)^r$ or $\left\|x - x^\dagger\right\|^r$ in Condition 6.29 in combination with Theorem 6.16 or Theorem 6.28 and (4.1) give the respective convergence rate results. □

6.7 Strict variational inequalities

Similar variational inequalities to the ones in Conditions 6.12 and 6.19 were considered also in [24]. In our notation and setup they would lead as follows.

Condition 6.32. (Strict VIE) Assume that for $x^\dagger \in \mathcal{L}$ there exist $\xi \in \partial\Psi(x^\dagger)$, $0 < \kappa \leq 1$, $\rho, \eta > 0$ and $\beta_i, \gamma_i \geq 0$, $i = 1, 2, 3$, such that

$$-\left\langle \xi, x - x^\dagger \right\rangle \leq \beta_1 \left(\Psi(x) - \psi^\dagger \right) + \beta_2 D_{\mathcal{T}}\left(x, x^\dagger\right) + \beta_3 \left\| F(x) - F(x^\dagger) \right\|^\kappa \tag{6.30}$$

$$\mathcal{T}\left(x, x^\dagger\right) \leq \gamma_1 \left(\Psi(x) - \psi^\dagger \right) + \gamma_2 D_{\mathcal{T}}\left(x, x^\dagger\right) + \gamma_3 \left\| F(x) - F(x^\dagger) \right\|^\kappa, \tag{6.31}$$

holds for all $x \in \tilde{V}_{\mathcal{L}}(\rho, \eta)$ as defined in (6.12) and the constants β_i, γ_i fulfill (6.11).

Condition 6.33. Assume that for $x^\dagger \in \mathcal{L}$, there exist $\xi \in \partial\Psi(x^\dagger)$, $\mu_i \geq 0$, $r, \rho > 0$, and $0 < \kappa \leq 1$ such that

$$\Psi(x - x^\dagger)^r \leq \mu_1 \left(\Psi(x) - \psi^\dagger \right) + \mu_2 D_{\mathcal{T}}\left(x, x^\dagger\right) + \mu_3 \left\| F(x) - F(x^\dagger) \right\|^\kappa \tag{6.32a}$$

or

$$\left\| x - x^\dagger \right\|^r \leq \mu_1 \left(\Psi(x) - \psi^\dagger \right) + \mu_2 D_{\mathcal{T}}\left(x, x^\dagger\right) + \mu_3 \left\| F(x) - F(x^\dagger) \right\|^\kappa \tag{6.32b}$$

holds for all $x \in \tilde{V}_{\mathcal{L}}(\rho, \eta)$ as defined in (6.12).

If α is chosen according to MDP we restrict our attention to the sets $V_{\mathcal{L}}(\rho)$ in (6.3) (or even subsets thereof, cf. Section 6.6), where Conditions 6.32 and 6.33 are stronger assumptions than Conditions 6.12 and 6.19, respectively, because

$$\Psi(x) - \psi^\dagger \leq 0 \leq D_\Psi^\xi\left(x, x^\dagger\right) \tag{6.33}$$

holds for $x \in V_{\mathcal{L}}(\rho)$. Therefore Theorems 6.16 and 6.21 remain applicable and we obtain the same convergence rates.

Note, however, that (6.30) is equivalent to the previous formulation (6.9) involving the Bregman distance. This can be seen as (6.30) implies (6.9) due to (6.33). The other implication follows since (6.9) implies that

$$-\left\langle \xi, x - x^\dagger \right\rangle \leq \beta_1' \left(\Psi(x) - \psi^\dagger \right) + \beta_2' D_{\mathcal{T}}\left(x, x^\dagger\right) + \beta_3' \left\| F(x) - F(x^\dagger) \right\|^\kappa.$$

where $\beta_i' = \beta_i/(1 - \beta_1)$ for $i = 1, 2, 3$, which is (6.30).

Corollary 6.34. *If Condition 6.32 is fulfilled for $\xi \in \partial\Psi(x^\dagger)$ and $\alpha = \alpha(\delta, y^\delta)$ is chosen according to MDP then for $x_\alpha^\delta \in \mathcal{M}_{\alpha, y^\delta}$ satisfying (4.1), it holds that*

$$D_\Psi^\xi\left(x_\alpha^\delta, x^\dagger\right) = \mathcal{O}(\delta^\kappa) \quad as \quad \delta \to 0, \tag{6.34}$$

$$D_{\mathcal{T}}\left(x_\alpha^\delta, x^\dagger\right) = \mathcal{O}(\delta^\kappa) \quad as \quad \delta \to 0. \tag{6.35}$$

If, moreover, Condition 6.33 is satisfied, then

$$\Psi(x_\alpha^\delta - x^\dagger) = \mathcal{O}(\delta^{\kappa/r}) \quad \text{or} \quad \left\|x_\alpha^\delta - x^\dagger\right\| = \mathcal{O}(\delta^{\kappa/r}),$$

as $\delta \to 0$ hold for (6.32a) or (6.32b), respectively.

Proof. Since $V_\mathcal{L}(\rho) \subset \tilde{V}_\mathcal{L}(\rho, \eta)$ for all $\eta > 0$ we obtain from (6.33) that Conditions 6.12 and 6.19 are fulfilled and thus Theorems 6.16 and 6.21 provide the results. □

We have seen that the variational inequalities in Condition 6.32 are stronger assumptions than the ones used in Condition 6.12. However, they have a number of interesting features, which we will describe now. The proofs are adapted from [24]. First of all, for differentiable penalty terms, the strict nonlinearity condition (6.31) readily implies the source condition (6.30).

Lemma 6.35. *Let $\Psi(x)$ be differentiable at $x^\dagger \in \mathcal{L}$ and $\xi = \Psi'(x^\dagger)$. If there exist $\varepsilon, \gamma_1 > 0$ and $\gamma_2, \gamma_3 \geq 0$ such that $B_\varepsilon(x^\dagger) \subset \mathcal{D}$ and*

$$\mathcal{T}(x, x^\dagger) \leq \gamma_1 \left(\Psi(x) - \psi^\dagger\right) + \gamma_2 D_\mathcal{T}\left(x, x^\dagger\right) + \gamma_3 \left\|F(x) - F(x^\dagger)\right\| \quad (6.36)$$

holds for all $x \in B_\varepsilon(x^\dagger)$, then there exist $\beta \geq 0$ such that

$$-\langle \xi, x - x^\dagger \rangle \leq \beta D_\mathcal{T}\left(x, x^\dagger\right) \quad (6.37)$$

holds for all $x \in X$.

Proof. We fix $z \neq 0$ and apply (6.36) to $z_t = x^\dagger + tz$ (which belongs to $B_\rho(x^\dagger)$ for $t > 0$ small enough). Dividing by t thus yields

$$\frac{1}{t}\left\|F(x^\dagger + tz) - F(x^\dagger) - F'(x^\dagger)(tz)\right\|$$

$$\leq \gamma_1 \frac{\Psi(x^\dagger + tz) - \Psi(x^\dagger)}{t} + \gamma_2 \left\|F'(x^\dagger)z\right\| + \gamma_3 \frac{1}{t}\left\|F(x^\dagger + tz) - F(x^\dagger)\right\|.$$

Taking the limit $t \to 0^+$ and choosing $z = x - x^\dagger$ for $x \in X \backslash \{x^\dagger\}$ (note, that if $x = x^\dagger$, then (6.37) is satisfied trivially) we obtain

$$0 \leq \gamma_1 \langle \Psi'(x^\dagger), x - x^\dagger \rangle + (\gamma_2 + \gamma_3)\left\|F'(x^\dagger)(x - x^\dagger)\right\|,$$

which implies (6.37) with

$$\beta = \frac{\gamma_2 + \gamma_3}{\gamma_1}.$$

□

Another interesting feature of the stricter variational inequality (6.32b) with $r = 2, \mu_3 = 0$ is that in Hilbert spaces it is equivalent to the classical range inclusion source condition

$$\operatorname{rg}(F'(x^\dagger)) \cap \partial \Psi(x^\dagger) \neq \emptyset$$

which in turn implies a variational source condition (6.30) (cf. Example 6.43 below).

Lemma 6.36. *Let X be a Hilbert space and let there exists $c > 0$ such that*

$$\|x - z\|^2 \leq c D_\Psi^\xi(x, z)$$

for all $x, z \in \mathcal{D}$, $\xi \in \partial \Psi(z) \neq \emptyset$. Then to $x^\dagger \in \mathcal{L}$ there exist $\mu_1, \mu_2 > 0$ such that

$$\left\|x - x^\dagger\right\|^2 \leq \mu_1 \left(\Psi(x) - \psi^\dagger\right) + \mu_2 D_\mathcal{T}\left(x, x^\dagger\right) \qquad (6.38)$$

holds for all $x \in X$, if and only if

$$\operatorname{rg}(F'(x^\dagger)^*) \cap \partial \Psi(x^\dagger) \neq \emptyset$$

Proof. Let $x^\dagger \in \mathcal{L}$ and $\xi = F'(x^\dagger)^* w \in \partial \Psi(x^\dagger)$ with $w \in Y^*$, then

$$\left\|x - x^\dagger\right\|^2 \leq c D_\Psi^\xi\left(x, x^\dagger\right)$$
$$\leq c\left(\Psi(x) - \psi^\dagger\right) - c\langle F'(x^\dagger)^* w, x - x^\dagger\rangle$$
$$\leq c\left(\Psi(x) - \psi^\dagger\right) + c \|w\|_{Y^*} D_\mathcal{T}\left(x, x^\dagger\right)$$

and we may choose $\mu_1 = c$ and $\mu_2 = c \|w\|_{Y^*}$ in (6.38).

On the other hand, to show the converse implication, we note that in (6.38) both sides of the inequality are convex. Indeed, Ψ is convex according to our standing assumptions in Condition 2.7 and one easily checks that the other terms involving the Hilbert space norm and the bounded, linear operator $F'(x^\dagger)$ are as well. Moreover, both sides vanish for $x = x^\dagger$. Therefore, their subderivatives exist and we obtain

$$\{0\} = \partial \left(\left\|. - x^\dagger\right\|^2\right)(x^\dagger) \subset \partial \Psi(x^\dagger) + \mu_2 \partial \left(D_\mathcal{T}\left(., x^\dagger\right)\right)(x^\dagger).$$

From the definition of $D_\mathcal{T}\left(x, x^\dagger\right)$ in (6.7) it thus follows that there exists

$$\tilde{w} \in \frac{\mu_2}{\mu_1} \partial \left(\left\|F'(x^\dagger)(. - x^\dagger)\right\|\right)(x^\dagger)$$

such that

$$0 \in \partial \Psi(x^\dagger) + F'(x^\dagger)^* \tilde{w}.$$

Choosing $w = -\tilde{w}$ we see that
$$\xi := F'(x^\dagger)^* w \in \partial \Psi(x^\dagger)$$
and, consequently, that
$$\mathrm{rg}(F'(x^\dagger)^*) \cap \partial \Psi(x^\dagger) \neq \emptyset.$$
$\square$

6.8 Convergence rates under a scaling invariant nonlinearity condition

Convergence rates with respect to Bregman distances for Tikhonov-type functionals with convex penalty terms have first been proven by Burger and Osher [9], who focused mainly on the case of linear operators, but also proposed the following nonlinear generalization of their results.

Condition 6.37. Let $x^\dagger \in \mathcal{L}$ as defined in (2.3) and assume that there is $w \in Y^*$ such that
$$\xi := F'(x^\dagger)^* w \in \partial \Psi(x^\dagger) \qquad (6.39)$$
and that there are $c, \rho > 0$ such that
$$\left\langle w, F(x) - F(x^\dagger) - F'(x^\dagger)(x - x^\dagger) \right\rangle \le c \, \|w\|_{Y^*} \left\| F(x) - F(x^\dagger) \right\|. \qquad (6.40)$$
holds for all $x \in V_\mathcal{L}(\rho)$ as defined in (6.3).

Using the above condition we are now ready to prove the same convergence rates for Morozov's discrepancy principle which were established in [9] for a-priori parameter choice rules. In the case of linear operators a similar result has been shown in [6].

Lemma 6.38. *If Condition 6.37 holds, then there exist $\beta > 0$ such that for all $x \in V_\mathcal{L}(\rho)$*
$$D_\Psi^\xi \left(x, x^\dagger \right) \le \beta \left\| F(x) - F(x^\dagger) \right\|$$
holds.

Proof. Since $\Psi(x) \le \psi^\dagger$ for $x \in V_\mathcal{L}(\rho)$, we obtain from (6.39) and (6.40) that
$$\begin{aligned}
D_\Psi^\xi \left(x, x^\dagger \right) &\le \Psi(x) - \Psi(x^\dagger) - \langle F'(x^\dagger)^* w, x - x^\dagger \rangle \\
&\le \left\langle w, F(x) - F(x^\dagger) - F'(x^\dagger)(x - x^\dagger) \right\rangle - \left\langle w, F(x) - F(x^\dagger) \right\rangle \\
&\le c \, \|w\|_{Y^*} \left\| F(x) - F(x^\dagger) \right\| + \left| \left\langle w, F(x) - F(x^\dagger) \right\rangle \right| \\
&\le \beta \left\| F(x) - F(x^\dagger) \right\|,
\end{aligned}$$

where $\beta = (c+1)\,\|w\|_{Y^*}$. □

As a consequence we obtain a convergence rate with respect to the Bregman distance when using MDP to choose the regularization parameter.

Theorem 6.39. *Let 6.37 hold for $x^\dagger \in \mathcal{L}$ and let ξ be as in (6.39). If $\alpha = \alpha(\delta, y^\delta)$ is chosen according to MDP then for $x_\alpha^\delta \in \mathcal{M}_{\alpha, y^\delta}$ satisfying (4.1), it holds that*

$$D_\Psi^\xi\left(x_\alpha^\delta, x^\dagger\right) = \mathcal{O}(\delta). \tag{6.41}$$

as $\delta \to 0$.

Proof. According to Lemma 6.3 we know that $x_\alpha^\delta \in V_\mathcal{L}(\rho)$ whenever δ is small enough. Thus, if Condition 6.37 holds, we apply Lemma 6.38 to obtain β such that

$$D_\Psi^\xi\left(x_\alpha^\delta, x^\dagger\right) \leq \beta\,\left\|F(x_\alpha^\delta) - F(x^\dagger)\right\| \leq \beta\tau_2\delta = \mathcal{O}(\delta),$$

where the last estimate follows from Lemma 4.4. □

Note that the nonlinearity condition (6.40) is not strong enough to ensure a convergence rate with respect to the Taylor-type distance $D_\mathcal{T}\left(x, x^\dagger\right)$ as we had in (6.19). Therefore, we have to omit this term in the variational inequalities on $\Psi(x - x^\dagger)$ or $\left\|x - x^\dagger\right\|$ in order to still obtain strong convergence rate results.

Theorem 6.40. *Let Conditions 6.37 be satisfied and assume that to $x^\dagger \in \mathcal{L}$ there exist $\xi \in \partial\Psi(x^\dagger)$, $\mu_1, \mu_3 \geq 0$ and $r, \rho > 0$ such that either*

$$\Psi(x - x^\dagger)^r \leq \mu_1 D_\Psi^\xi\left(x, x^\dagger\right) + \mu_3 \left\|F(x) - F(x^\dagger)\right\| \tag{6.42a}$$

or

$$\left\|x - x^\dagger\right\|^r \leq \mu_1 D_\Psi^\xi\left(x, x^\dagger\right) + \mu_3 \left\|F(x) - F(x^\dagger)\right\| \tag{6.42b}$$

for all $x \in V_\mathcal{L}(\rho)$. If $\alpha = \alpha(\delta, y^\delta)$ is chosen according to MDP then for $x_\alpha^\delta \in \mathcal{M}_{\alpha, y^\delta}$ satisfying (4.1)

$$\Psi(x_\alpha^\delta - x^\dagger) = \mathcal{O}(\delta^{1/r}) \quad \text{or} \quad \left\|x_\alpha^\delta - x^\dagger\right\| = \mathcal{O}(\delta^{1/r}),$$

as $\delta \to 0$ hold for (6.42a) or (6.42b), respectively.

Proof. The convergence rates follow since

$$\mu_1 D_\Psi^\xi\left(x_\alpha^\delta, x^\dagger\right) + \mu_3 \left\|F(x_\alpha^\delta) - F(x^\dagger)\right\| = \mathcal{O}(\delta)$$

according to Theorem 6.39 and Lemma 4.4 from (6.42a) or (6.42b), respectively. □

6.9 The linear case

We now summarize our findings for the special case of bounded, linear operators F, where (6.10) becomes a tautology and

$$D_\mathcal{T}(x, x^\dagger) = \left\| F(x) - F(x^\dagger) \right\|.$$

Without loss of generality we may thus assume $\beta_2 = \gamma_i = \mu_2 = 0$ for $i = 1, 2, 3$ in Condition 6.12.

Corollary 6.41. *Assume that F is a bounded, linear operator and that for $x^\dagger \in \mathcal{L}$ there exist $\xi \in \partial\Psi(x^\dagger)$, $0 < \kappa \leq 1$ and $\rho > 0$ such that*

$$-\langle \xi, x - x^\dagger \rangle \leq \beta_1 D_\Psi^\xi(x, x^\dagger) + \beta_3 \left\| F(x) - F(x^\dagger) \right\|^\kappa$$

holds for all $x \in V_\mathcal{L}(\rho)$ with $0 \leq \beta_1 < 1$ and $\beta_3 \geq 0$. Then, if $\alpha = \alpha(\delta, y^\delta)$ is chosen according to MDP and $x_\alpha^\delta \in \mathcal{M}_{\alpha, y^\delta}$ satisfies (4.1), we have

$$D_\Psi^\xi(x_\alpha^\delta, x^\dagger) = \mathcal{O}(\delta^\kappa)$$

as $\delta \to 0$. If, furthermore, there exist $\mu_1, \mu_3 \geq 0$ and $r > 0$ such that either

$$\Psi(x - x^\dagger)^r \leq \mu_1 D_\Psi^\xi(x, x^\dagger) + \mu_3 \left\| F(x) - F(x^\dagger) \right\|^\kappa,$$

or

$$\left\| x - x^\dagger \right\|^r \leq \mu_1 D_\Psi^\xi(x, x^\dagger) + \mu_3 \left\| F(x) - F(x^\dagger) \right\|^\kappa,$$

holds for all $x \in V_\mathcal{L}(\rho)$, then

$$\Psi(x_\alpha^\delta - x^\dagger) = \mathcal{O}(\delta^{\kappa/r}) \quad \text{or} \quad \left\| x_\alpha^\delta - x^\dagger \right\| = \mathcal{O}(\delta^{\kappa/r}),$$

as $\delta \to 0$, respectively.

Proof. Since Conditions 6.12 and 6.19 are satisfied, we obtain the respective results from Theorems 6.16 and 6.21. □

Remark 6.42. Also, or maybe especially, for linear operators one can identify conditions such that localized variational inequalities as in Section 6.6 suffice. If F is injective or if Ψ is strictly convex, then in Condition 6.30 $\mathcal{P} = \{x^\dagger\} = \mathcal{L}$. Therefore, the combination of Conditions 6.29 and 6.12 or – if Ψ additionally satisfies Condition 3.7 – 6.24, ensures that Theorem 6.31 is applicable. ∎

6.10 Relation of variational inequalities to source and nonlinearity conditions

Variational formulations of source and nonlinearity conditions have been used earlier in order to obtain convergence rate results. In this section, we would like to draw a connection between inequalities (6.9) and (6.10) in Condition 6.12 and classical source and nonlinearity conditions. Variational inequalities can be seen as a generalization of the latter as the following examples illustrate.

Example 6.43.   (i) If $\xi \in \partial\Psi(x^\dagger)$ fulfills the classical source condition

$$\xi = F'(x^\dagger)^* w, \qquad (6.43)$$

with $w \in Y^*$, then it follows that

$$-\langle \xi, x - x^\dagger \rangle \leq \left|\langle w, F'(x^\dagger)(x - x^\dagger)\rangle\right| \leq \|w\|_{Y^*} D_{\mathcal{T}}(x, x^\dagger), \qquad (6.44)$$

and thus (6.9) holds with $\beta_2 = \|w\|_{Y^*}$, and $\beta_1 = \beta_3 = 0$. Note, that the presence of the term $D_{\mathcal{T}}(x, x^\dagger)$ in (6.9) allows us to express this classical source condition through only the first variational inequality. Omitting this term and using an alternative formulation

$$-\langle \xi, x - x^\dagger \rangle \leq \beta_1 D_\Psi^\xi(x, x^\dagger) + \beta_3 \left\|F(x) - F(x^\dagger)\right\|^\kappa, \qquad (6.45)$$

which has been considered, e.g., in [5, 56], one always needs to also employ some sort of structural nonlinearity condition to include this standard case in the setting. Nevertheless, (6.45) combined with a nonlinearity condition (such as (6.10)) is equivalent to Condition 6.12 in the sense of Lemma 6.14.

(ii) One of the first structural assumptions regarding nonlinearity (see, e.g., [17, 48]), was that F be Fréchet differentiable between Hilbert spaces X, Y and that the derivative be Lipschitz continuous, i.e., for some $\rho > 0$ and all $x, z \in X$ it holds

$$\left\|F'(x) - F'(z)\right\| \leq c \left\|x - z\right\|.$$

Under this assumption one can show that for classical Tikhonov regularization, where $\Psi(x) = \|x\|^2$, the following estimate holds

$$\mathcal{T}(x, x^\dagger) \leq \frac{c}{2} \left\|x - x^\dagger\right\|^2 = \frac{c}{2} D_\Psi^\xi(x, x^\dagger),$$

which is (6.10) with $\gamma_1 = c/2$ and $\gamma_2 = \gamma_3 = 0$. In [52] the variational inequality

$$\mathcal{T}(x, x^\dagger) \leq c\, D_\Psi^\xi(x, x^\dagger). \qquad (6.46)$$

has been used as the nonlinearity condition for regularization with more general, convex penalty terms.

(iii) In [34] an operator F is defined to be *nonlinear of degree* (n_1, n_2, n_3) locally near $x^\dagger$, with $n_1, n_2 \in [0,1], n_3 \in [0,2]$, if for some $c, \rho > 0$ and all $x \in V_\mathcal{L}(\rho)$:

$$\mathcal{T}(x, x^\dagger) \leq c D_\mathcal{T}(x, x^\dagger)^{n_1} \left\|F(x) - F(x^\dagger)\right\|^{n_2} \left\|x - x^\dagger\right\|^{n_3}.$$

Taking into account the problem under consideration in [34], where X,Y are Hilbert spaces and $\Psi = \|.\|^2$, this definition may be generalized within our framework to

$$\mathcal{T}(x, x^\dagger) \leq c D_\mathcal{T}(x, x^\dagger)^{n_1} \left\|F(x) - F(x^\dagger)\right\|^{n_2} D_\Psi^\xi(x, x^\dagger)^{n_3}.$$

Applying Young's inequality twice to that last inequality we find that – whenever $n_1 + n_3 < 1$ – there exist constants γ_i such that (6.10) holds for x sufficiently close to $x^\dagger$ with

$$\kappa = \min\left(1, \frac{n_2}{1 - n_1 - n_3}\right).$$

Note that if one considers examples (i) and (ii) together for $\beta_1 = \gamma_2 = 0$ (cf. Lemma 6.14), then (6.11) becomes the well-known smallness condition

$$\frac{c}{2} \|w\|_{Y^*} < 1.$$

Chapter 7

Sparse recovery

7.1 Assumptions and problem formulation

As a prominent case study, we will show now that the convergence rate from (6.25) for the sparsity promoting penalty terms $\Psi_{p,w}$ defined in (2.9) can be improved significantly when including the a-priori information that the $\Psi_{p,w}$-minimizing solution $x^\dagger$ is also sparse. To this end we show that a variational inequality as in Condition 6.19 holds true in this setting.

Condition 7.1. Let X be a Hilbert space and $\{\phi_\lambda\}_{\lambda \in \Lambda}$ be a fixed frame for X (cf. Definition 2.16).

For elements $x \in X$ we use the shorthand notation

$$x_\lambda = \langle \phi_\lambda, x \rangle.$$

Throughout this section we assume that $x^\dagger \in \mathcal{L}$ is sparse with respect to $\{\phi_\lambda\}_{\lambda \in \Lambda}$, i.e., that only finitely many $x^\dagger_\lambda$ are nonzero. For any $\xi \in \partial \Psi(x^\dagger) \subset X^* = X$ the sequence $\{\xi_\lambda\}_{\lambda \in \Lambda}$ belongs to $\ell_2(\Lambda)$ and thus the set

$$J = \{\lambda \in \Lambda \mid x^\dagger_\lambda \neq 0 \vee |\xi_\lambda| \geq w_0\} \tag{7.1}$$

is also finite. We denote the subspace spanned by elements with indices in J by

$$U = \operatorname{span}\{\phi_\lambda \mid \lambda \in J\},$$

and the projections of X onto U and $U^\perp$ by π and $\pi^\perp$, respectively.

Moreover, we assume that the restriction $F'(x^\dagger)|_U$ of $F'(x^\dagger)$ to U is injective, i.e., for all $x, z \in X$ from $F'(x^\dagger)(x - z) = \pi^\perp(x - z) = 0$ it follows that $x = z$.

We remark that $F'(x^\dagger)|_U$ is clearly injective, if $F'(x^\dagger)$ satisfies the so-called *FBI* property (see, e.g., [42]).

Definition 7.2. A bounded, linear operator $A : X \to Y$ satisfies the *Finite Basis Injectivity* (FBI) property, if for any finite dimensional subspace X_n of X the restriction $A|_{X_n}$ of A to X_n is injective.

7.2 Variational Inequalities

A variational inequality of type (6.20a) indeed holds in the sparse recovery case which will allow us to derive convergence rates whenever Condion 6.12 is satisfied. This is shown in Theorem 7.5 and Corollary 7.6 below. The proof is based on techniques from [24] and uses the following technical Lemma.

Lemma 7.3. *If condition 7.1 is satisfied and $\Psi_{p,w}$ is as in Definition 2.17, then there exists $c > 0$ such that for any $x \in X, \xi \in \partial \Psi_{p,w}(x^\dagger)$ and $\lambda \notin J$*

$$w_\lambda |x_\lambda|^p \leq c \left(w_\lambda |x_\lambda|^p - \xi_\lambda x_\lambda \right) \qquad (7.2)$$

holds.

Proof. We first consider the case $p > 1$, then the unique element in the subgradient $\xi \in \partial \Psi_{p,w}(x^\dagger)$,

$$\xi = \sum_{\lambda \in \Lambda} p\, w_\lambda \operatorname{sign}(\langle \phi_\lambda, x^\dagger \rangle) \left| \langle \phi_\lambda, x^\dagger \rangle \right|^{p-1} \phi_\lambda,$$

satisfies $\xi_\lambda = 0$ whenever $x_\lambda^\dagger = 0$, which in turn holds for all $\lambda \notin J$, so that (7.2) clearly holds for all $c \geq 1$.

On the other hand, for $p = 1$ we define

$$m = \max_{\lambda \notin J} |\xi_\lambda| < w_0.$$

Here the maximum is attained since $\xi \in X^* = X$ and thus the sequence $\{\xi_\lambda\}_{\lambda \in \Lambda}$ belongs to $\ell_2(\Lambda)$. Since $0 \leq |\xi_\lambda| \leq m < w_0 \leq w_\lambda$ the choice $c = 2w_0/(w_0 - m)$ yields

$$\frac{w_\lambda}{c} |x_\lambda| = w_\lambda |x_\lambda| - \frac{w_\lambda}{w_0} \frac{w_0 + m}{2} |x_\lambda|$$
$$\leq w_\lambda |x_\lambda| - m |x_\lambda| \leq w_\lambda |x_\lambda| - \xi_\lambda x_\lambda,$$

and (7.2) follows. □

We now show that a variational inequality (6.20a) holds in the sparse recovery case.

Lemma 7.4. *If condition 7.1 is satisfied, then there exist $\mu_1, \mu_2 > 0$ such that for all $x \in X$*

$$\Psi_{p,w}(x - x^\dagger) \leq \mu_1 D_{\Psi_{p,w}}^\xi(x, x^\dagger) + \mu_2 D_\mathcal{T}(x, x^\dagger)^p, \qquad 1 \leq p \leq 2. \qquad (7.3)$$

holds.

Proof. In order to estimate the difference between x and $x^\dagger$ with respect to the penalty term, we use the splitting

$$\Psi_{p,w}(x - x^\dagger) = \sum_{\lambda \in J} w_\lambda \left|x_\lambda - x_\lambda^\dagger\right|^p + \sum_{\lambda \notin J} w_\lambda \left|x_\lambda - x_\lambda^\dagger\right|^p, \qquad (7.4)$$

and write

$$c_w = \sup_{\lambda \in J}\{w_\lambda\},$$

which is a finite number because the set J, defined in (7.1), is finite. Using the equivalence of norms on finite dimensional spaces, we find a constant c_p such that

$$\sum_{\lambda \in J} w_\lambda \left|x_\lambda - x_\lambda^\dagger\right|^p \leq c_w \left\|\{x_\lambda - x_\lambda^\dagger\}_{\lambda \in J}\right\|^p_{\ell_p(J)}$$

$$\leq c_w c_p \left\|\{x_\lambda - x_\lambda^\dagger\}_{\lambda \in J}\right\|^p_{\ell_2(J)}$$

$$= c_w c_p \left\|\pi(x - x^\dagger)\right\|^p$$

Due to the injectivity of $F'(x^\dagger)$ on U, the boundedness of $F'(x^\dagger)$ and the inequality $(a + b)^p \leq 2(a^p + b^p)$ for $a, b \geq 0$ and $p \geq 1$, we get the following estimate.

$$\left\|\pi(x - x^\dagger)\right\|^p \leq c' \left\|F'(x^\dagger)\pi(x - x^\dagger)\right\|^p$$

$$\leq 2c' \left(\left\|F'(x^\dagger)(x - x^\dagger)\right\|^p + \left\|F'(x^\dagger)\right\|^p \left\|\pi^\perp x\right\|^p \right).$$

From the well known inequality $\|.\|_{\ell_2} \leq \|.\|_{\ell_p}$ for $1 \leq p \leq 2$, Lemma 7.3 and $x_\lambda^\dagger = 0$ for all $\lambda \notin J$, it follows that

$$\left\|\pi^\perp x\right\|^p = \left(\sum_{\lambda \notin J} |x_\lambda|^2\right)^{p/2} \leq \sum_{\lambda \notin J} \frac{w_\lambda}{w_0} |x_\lambda|^p$$

$$= \frac{c}{w_0} \sum_{\lambda \notin J} w_\lambda |x_\lambda|^p - w_\lambda \left|x_\lambda^\dagger\right|^p - \xi_\lambda(x_\lambda - x_\lambda^\dagger)$$

$$\leq \frac{c}{w_0} D^\xi_{\Psi_{p,w}}(x, x^\dagger),$$

where the last inequality holds because all remaining summands for $\lambda \in J$ are Bregman distances $D^{\xi_\lambda}_{w_\lambda |.|}(x_\lambda, x_\lambda^\dagger)$, where $\xi_\lambda \in \partial(w_\lambda |.|)(x_\lambda^\dagger)$, and hence nonnegative.

To obtain the remaining estimates for terms corresponding to $\lambda \notin J$ in (7.4), we again use Lemma 7.3 and $x_\lambda^\dagger = 0$ for all $\lambda \notin J$.

$$\sum_{\lambda \notin J} w_\lambda \left|x_\lambda - x_\lambda^\dagger\right|^p \leq c \sum_{\lambda \notin J} w_\lambda |x_\lambda|^p - w_\lambda \left|x_\lambda^\dagger\right|^p - \xi_\lambda(x_\lambda - x_\lambda^\dagger)$$

$$\leq c \, D^\xi_{\Psi_{p,w}}(x, x^\dagger).$$

Finally, collecting the above inequalities we find that

$$\Psi_{p,w}(x - x^\dagger) = \sum_{\lambda \notin J} w_\lambda \left|x_\lambda - x_\lambda^\dagger\right|^p + \sum_{\lambda \in J} w_\lambda \left|x_\lambda - x_\lambda^\dagger\right|^p$$

$$\leq \mu_1 \, D_{\Psi_{p,w}}^\xi(x, x^\dagger) + \mu_2 \, \left\|F'(x^\dagger)(x - x^\dagger)\right\|^p$$

holds for all $x \in X$, where

$$\mu_1 = 2c' c_w c_p \left\|F'(x^\dagger)\right\|^p \frac{c}{w_0} + c \quad \text{and} \quad \mu_2 = 2c'.$$

□

Theorem 7.5. *If Condition 7.1 is satisfied, then for $\varepsilon < \left\|F'(x^\dagger)\right\|^{-1}$ and $x \in B_\varepsilon(x^\dagger)$ it holds that*

$$\Psi_{p,w}(x - x^\dagger) \leq \mu_1 D_{\Psi_{p,w}}^\xi(x, x^\dagger) + \mu_2 D_{\mathcal{T}}(x, x^\dagger)^p, \quad 1 \leq p \leq 2, \quad (7.5)$$

with μ_1, μ_2 as in Lemma 7.4. Moreover, if additionally Condition 6.12 holds for arbitrary $\rho > 0$, then there exist $\mu_1, \mu_2 > 0$ such that (7.5) holds for all $x \in V_\rho(x^\dagger)$ as defined in (6.3).

Proof. The assumptions of Lemma 7.4 are fulfilled and using (7.3) and that $D_{\mathcal{T}}(x, x^\dagger) \leq 1$ whenever $x \in B_\varepsilon(x^\dagger)$, we find that

$$\Psi_{p,w}(x - x^\dagger) \leq \mu_1 \, D_{\Psi_{p,w}}^\xi(x, x^\dagger) + \mu_2 \, D_{\mathcal{T}}(x, x^\dagger)^p$$

$$\leq \mu_1 D_{\Psi_{p,w}}^\xi(x, x^\dagger) + \mu_2 \, D_{\mathcal{T}}(x, x^\dagger).$$

holds for all $x \in B_\varepsilon(x^\dagger)$. If, on the other hand, Condition 6.12 holds and $x \in V_\rho(x^\dagger)$, then

$$D_{\mathcal{T}}(x, x^\dagger) \leq \mathcal{T}(x, x^\dagger) + \left\|F(x) - F(x^\dagger)\right\| \leq \mathcal{T}(x, x^\dagger) + \rho$$

and due to (6.10), where $\gamma_2 < 1$ and $\Psi(x) \leq \Psi(x^\dagger)$, we therefore find that

$$D_{\mathcal{T}}(x, x^\dagger) \leq -\frac{\gamma_1}{1 - \gamma_2} \langle \xi, x - x^\dagger \rangle + \frac{\gamma_3}{1 - \gamma_2} \rho^\kappa + \frac{1}{1 - \gamma_2} \rho$$

Similarly, from (6.9) we obtain

$$-\langle \xi, x - x^\dagger \rangle \leq \frac{\beta_2}{1 - \beta_1} D_{\mathcal{T}}(x, x^\dagger) + \frac{\beta_3}{1 - \beta_1} \rho^\kappa,$$

so that together $D_{\mathcal{T}}(x, x^\dagger)$ is bounded by

$$D_{\mathcal{T}}(x, x^\dagger) \leq \left(1 - \frac{\beta_2 \gamma_1}{(1 - \beta_1)(1 - \gamma_2)}\right)^{-1} \left(\frac{\beta_3 \gamma_1 + \gamma_3(1 - \beta_1)}{(1 - \beta_1)(1 - \gamma_2)} \rho^\kappa + \frac{1}{1 - \gamma_2} \rho\right)$$

$$=: C.$$

Using (7.3) we find that for $x \in V_\rho(x^\dagger)$

$$\Psi_{p,w}(x - x^\dagger) \leq \mu_1 D^\xi_{\Psi_{p,w}}(x, x^\dagger) + \mu_2 D_T(x, x^\dagger)^p$$
$$\leq \mu_1 D^\xi_{\Psi_{p,w}}(x, x^\dagger) + \mu_2 \max(1, C^p) D_T(x, x^\dagger).$$

□

7.3 Convergence rates

As a corollary we obtain the convergence rate result.

Corollary 7.6. *If $x^\dagger \in \mathcal{L}$ satisfies Condition 6.12 and 7.1 and $\alpha = \alpha(\delta, y^\delta)$ is chosen according to MDP, then for $x^\delta_\alpha \in \mathcal{M}_{\alpha, y^\delta}$ satisfying (4.1) we obtain a convergence rate*

$$\Psi_{p,w}(x^\delta_\alpha - x^\dagger) = \mathcal{O}(\delta^\kappa) \quad as \quad \delta \to 0. \tag{7.6}$$

Proof. According to Theorem 7.5 we find that Condition 6.19 is satisfied with $r = 1$ and thus Theorem 6.21 is applicable and provides the result.
□

If we take X to be the sequence space ℓ_2 with the canonical basis, then the penalty terms $\Psi_{p,w}$ are powers of the weighted ℓ_p-norms, namely

$$\Psi_{p,w}(x) = \|x\|^p_{p,w}$$

and therefore (7.6) corresponds to a convergence rate

$$\left\| x^\delta_\alpha - x^\dagger \right\|_{p,w} = \mathcal{O}(\delta^{\kappa/p})$$

and if $\kappa = 1$ we obtain linear convergence speed for ℓ_1-regularization (compare [24, 25]).

Chapter 8

Autoconvolution

To illustrate the theoretical results of the previous sections, we will analyze a specific example that meets the imposed conditions, namely the autoconvolution operator over a finite interval.

This operator is of importance, for example, in stochastics, where it describes the density of the sum of two independent and identically distributed random variables, or also in spectroscopy (see [21] and the references therein for further details).

8.1 Properties

Let $X = Y = L^2[0,1]$, and define for $f \in \text{dom}(F) \subset X$ and $s \in [0,1]$ the operator $F : \text{dom}(F) \subset X \to Y$ through

$$F(f)(s) = (f * f)(s) = \int_0^s f(s-t)f(t)dt. \tag{8.1}$$

Autoconvolution has been studied in some detail in [21], where the authors showed that for the choice

$$\text{dom}(F) = D^+ := \{f \in L^2[0,1] : f(t) \geq 0 \quad \text{a.e. in} \quad [0,1]\}$$

the operator is weakly continuous and since D^+ is weakly closed in $L^2[0,1]$, also weakly sequentially closed.

The Fréchet derivative of F at the point $f \in X$ is given by the bounded, linear operator $F'(f) : X \to X$ defined as

$$[F'(f)h](s) = 2\int_0^s f(s-t)h(t)dt \qquad 0 \leq s \leq 1.$$

Indeed, we have

$$F(f+h) \quad F(f) - F'(f)h = F(h)$$

and therefore
$$\left\|F(f+h) - F(f) - F'(f)h\right\| = \|F(h)\| \leq \|h\|^2,$$
where $\|\cdot\|$ denotes the Hilbert space norm on $L^2[0,1]$, and the last inequality holds since for all $f, h \in X$
$$\|f * h\| \leq \|f\| \, \|h\|. \quad (8.2)$$
For a proof see [21, Theorem 2, Lemma 4]. Moreover, $F'(\cdot)$ is linear and due to (8.2)
$$\left\|F'(f-g)\right\| = \sup_{\|h\|=1} \left\|F'(f-g)h\right\| = \sup_{\|h\|=1} \|2(f-g)*h\| \leq 2\,\|f-g\|$$
holds, which shows that F' is Lipschitz continuous.

The adjoint of the Fréchet derivative $F'(f)^*h$ for $f, h \in X$ evaluates to
$$\langle F'(f)v, h \rangle = \int_0^1 2 \int_0^s f(s-t)v(t)dt\, h(s)ds$$
$$= \int_0^1 v(t) \int_t^1 2f(s-t)h(s)ds\, dt$$
$$= \langle v, F'(f) * h \rangle.$$
Writing $\tilde{h}(t) = (h)\tilde{\,}(t) = h(1-t)$ for any $h \in X$ we get
$$[F'(f)^*h](t) = 2 \int_t^1 f(s-t)\,h(s)\,ds$$
$$= 2 \int_0^{1-t} f(s)\,\tilde{h}((1-t)-s)\,ds$$
$$= 2\,(f * \tilde{h})\,(1-t) = 2\,(f * \tilde{h})\tilde{\,}(t). \quad (8.3)$$

8.2 Formulation in wavelet coefficients

In this example we will be reconstructing a solution f of
$$F(f) = g \quad (8.4)$$
from given noisy data g^δ, $\|g - g^\delta\| \leq \delta$. Among all possible solutions f we are interested in choosing the one with the sparsest representation in the Haar wavelets, which form an orthonormal basis of the Hilbert space X. Finding sparse solutions using wavelets is commonly used, e.g., in signal compression, but can also be of importance to obtain a good resolution of discontinuities in the solution. We define
$$\varphi(t) = \begin{cases} 1 & \text{if } 0 \leq t < 1/2 \\ -1 & \text{if } 1/2 \leq t < 1 \end{cases}$$
$$\psi(t) = 1 \quad 0 \leq t < 1$$

and for $j \in \mathbb{N}_0$ and $k \in \{0, \ldots, 2^j - 1\}$
$$\varphi_{j,k}(t) = 2^{j/2} \varphi(2^j t - k).$$
The coefficient vector of $f \in X$ in the Haar wavelet basis will be denoted by $x = \{x_\lambda\}_{\lambda \in \Lambda}$, where
$$\Lambda = \{1\} \cup \{(j,k) : j \in \mathbb{N}_0, 0 \leq k \leq 2^j - 1\},$$
$$\varphi_\lambda = \begin{cases} \psi & \text{if } \lambda = 1 \\ \varphi_{j,k} & \text{if } \lambda = (j,k) \end{cases}$$
and
$$x_\lambda = \langle f, \varphi_\lambda \rangle \quad \forall \lambda \in \Lambda$$
We now reformulate the problem in the coefficient space, so that all computations can be executed only on sequences in ℓ_2. To this end we express the operator F as follows,
$$F(f)(s) = F(\sum_{\lambda \in \Lambda} x_\lambda \varphi_\lambda)(s)$$
$$= \int_0^s \sum_{\lambda \in \Lambda} x_\lambda \varphi_\lambda(s-t) \sum_{\mu \in \Lambda} x_\mu \varphi_\mu(t) dt$$
$$= \sum_{\lambda, \mu \in \Lambda} x_\lambda x_\mu \, \varphi_\lambda * \varphi_\mu(s).$$
Since $F(f) \in Y = X$ we can also represent the image of f with respect to the Haar wavelet basis and obtain
$$y_\eta = \langle F(f), \varphi_\eta \rangle = \sum_{\lambda, \mu \in \Lambda} x_\lambda x_\mu \, \langle \varphi_\lambda * \varphi_\mu, \varphi_\eta \rangle = x^T K_\eta \, x, \qquad (8.5)$$
where
$$K_\eta := \left\{ \langle \varphi_\lambda * \varphi_\mu, \varphi_\eta \rangle \right\}_{\lambda, \mu \in \Lambda} \quad \forall \eta \in \Lambda. \qquad (8.6)$$
The above representation allows us to consider $\tilde{F} : \ell_2 \to \ell_2$, with
$$\tilde{F}(x) = \{y_\eta\}_{\eta \in \Lambda}, \quad y_\eta := x^T K_\eta \, x. \qquad (8.7)$$

For the iterative computation of the regularized solutions in (8.8) below, we will also need to express the Fréchet derivative of F in the coefficient space. Adapting the steps leading up to (8.5) we obtain from (8.3)
$$\langle F'(f)^* h, \varphi_\eta \rangle = 2 \, (x^T K_\eta \, \tilde{z}),$$
where the matrices K_η were defined in (8.6) and $\tilde{z}$ denotes the coefficients of $\tilde{h}(t) = h(1-t)$. They can be computed according to the formulae
$$\tilde{z}_1 = \langle \tilde{h}, \psi \rangle = \langle h, \tilde{\psi} \rangle = \langle h, \psi \rangle = z_1,$$
$$\tilde{z}_{(j,k)} = \langle \tilde{h}, \varphi_{(j,k)} \rangle = \langle h, \tilde{\varphi}_{(j,k)} \rangle = \langle h, -\varphi_{(j,k)'} \rangle = -z_{(j,k)'},$$
where $(j,k)' := (j, 2^j - 1 - k)$.

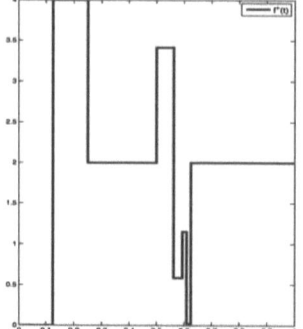 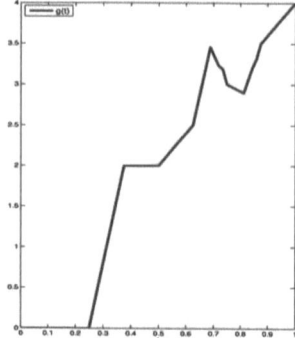

Figure 8.1: Plot of the Ψ-minimizing solution $f^\dagger$ (left), and the exact data $g = F(f^\dagger)$ (right).

8.3 Numerical results

In order to find a solution that has a sparse representation, we choose the penalty term to be $\Psi(x) = \|x\|_1$ as it is well known that regularization with the ℓ_1 norm promotes sparsity (see, e.g., [13, 24, 38, 48, 50]). We thus consider the Tikhonov-type functionals

$$J_{\alpha,y^\delta}(x) = \left\|\tilde{F}(x) - y^\delta\right\|^2 + \alpha \|x\|_1,$$

where from now on y, y^δ denote the coefficient vectors of the exact data g and the noisy data g^δ, respectively. In this notation it holds that $\|g - g^\delta\| = \|y - y^\delta\|$ and therefore the condition for the noise level can be equivalently stated in the ℓ_2 framework. It simply reads $\|y - y^\delta\| \leq \delta$.

To compute the regularized solutions

$$x_\alpha^\delta = \arg\min_{x \in \ell_2} J_{\alpha,y^\delta}(x)$$

we will use the iterative soft-shrinkage algorithm for non-linear inverse problems from [48], which is based on the surrogate functional approach described in [13]. We denote the soft-shrinkage operator with threshold $\beta > 0$ by S_β, i.e. for $x = \{x_\lambda\}_{\lambda \in \Lambda} \in \ell_2$

$$(S_\beta(x))_\lambda = \begin{cases} x_\lambda - \beta & \text{if } x_\lambda > \beta \\ x_\lambda + \beta & \text{if } x_\lambda < -\beta \\ 0 & \text{if } |x_\lambda| \leq \beta. \end{cases}$$

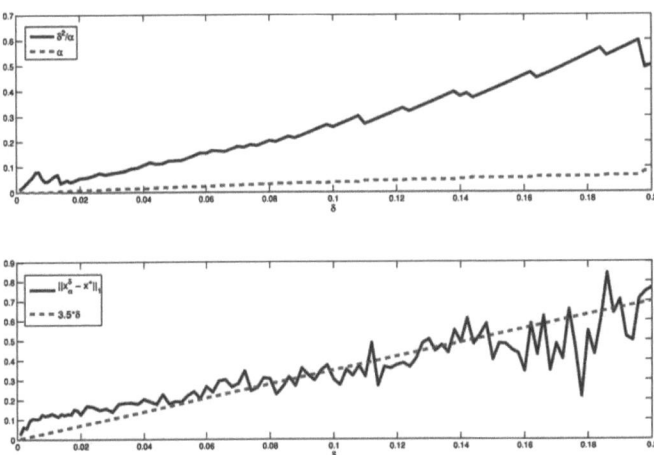

Figure 8.2: Top: Graph of $\delta \mapsto \delta^2/\alpha(\delta, y^\delta)$ (solid) and of $\delta \mapsto \alpha(\delta, y^\delta)$ (dashed). Bottom: Graph of $\delta \mapsto \left\| x^\delta_{\alpha(\delta,y^\delta)} - x^\dagger \right\|_1$ with the straight line $\delta \mapsto 3.5 \cdot \delta$ for comparison. The noise was chosen uniformly random at each step.

It has been shown in [48] that for arbitrary $x^0 \in \ell_2$ and with

$$x^{n,0} = x_n,$$
$$x^{n,k+1} = S_{\alpha/2C}\left(x^{n,0} + \frac{1}{C}F'(x^{n,k})^*\left(y^\delta - F(x^{n,0})\right)\right), \quad (8.8)$$
$$x^{n+1} = \lim_{k \to \infty} x^{n,k}$$

the resulting sequence x_n converges at least to a critical point of $J_{\alpha,y^\delta}(x)$. Here the constant C has to be choosen large enough for the algorithm to converge (see [48] for details). Choosing the starting value x^0 reasonably close to the true solution we observed numerically that in this case the iteration actually approximates a minimizer.

Moreover, choosing the right hand side g such that

$$g \in R_\varepsilon^+ := \{g \in C[0,1] \ : \ g \geq 0, \ \varepsilon = \max\{s : g(\xi) = 0 \ \forall \xi \in [0,s]\}\}$$

for some $\varepsilon > 0$, it has been shown in [21] that any $f \in D(F)$ fulfilling (8.4) possesses the form

$$f(t) = \begin{cases} 0 & \text{a.e. in } t \in [0, \varepsilon/2] \\ \text{uniquely determined by } g & \text{a.e. in } t \in [\varepsilon/2, 1 - \varepsilon/2] \\ \text{arbitrarily non-negative} & \text{in } t \in [1 - \varepsilon/2, 1] \end{cases} \quad (8.9)$$

For our test, we have chosen $g \in R_\varepsilon^+$ with $\varepsilon = 1/4$ as shown in Figure 8.1. Due to the representation of the possible solutions in (8.9), we know that they are uniquely determined on $[0, \frac{7}{8}]$, but differ on $(\frac{7}{8}, 1]$. Also, the right hand side was chosen such that it actually permits a sparse solution. The Ψ-minimizing solution $f^\dagger$ (see Figure 8.1) has four non-zero coefficients when expressed in the Haar wavelet basis. Note that in this example the solution $f^\dagger$ is different from the minimum norm solution, which would have the constant value zero on $(\frac{7}{8}, 1]$.

Since we have seen that the operator F and the penalty term Ψ fulfill all the conditions needed to apply Theorem 5.9 and Theorem 3.10 (a parameter $\alpha(\delta, y^\delta)$ could indeed be found for all choices of the noisy data, ensuring that the discrepancy principle is applicable), and since moreover the Ψ-minimizing solution $x^\dagger$ is unique, it can be expected that for $\delta \to 0$ we observe

$$\alpha(\delta, y^\delta) \to 0, \quad \frac{\delta^2}{\alpha(\delta, y^\delta)} \to 0, \quad \text{and} \quad \Psi(x_\alpha^\delta - x^\dagger) = \left\| x_\alpha^\delta - x^\dagger \right\|_1 \to 0.$$

Our numerical experiments confirm these expectations. For a sample case with wavelets up to a maximal index level $J = 5$ and noise chosen uniformly random such that $\left\| y - y^\delta \right\| = \delta$, $\delta \in (0, 0.2]$ at each step, we show the results in Figure 8.2. For comparison we have added the straight line with slope 3.5. The numeric results thus visually suggest that a convergence rate of order δ is obtained in the topology induced by the ℓ_1-penalty term. Due to our results on sparse recovery in Chapter 7 this comes as no surprise. However, to the best of our knowledge results on a variational non-linearity condition (6.10) are not available in the literature for the autoconvolution operator.

The corresponding values of $\alpha = \alpha(\delta, y^\delta)$ were found using Algorithm 1 in Section 5.4 which is based on the monotonicity of the residual functional $g(\alpha)$ (cf. Lemma 4.7). The reconstructed solutions have between three and seven non-zero coefficients for $10^{-2} \leq \delta \leq 0.2$ and eight to eleven non-zero coefficients for $10^{-3} \leq \delta < 10^{-2}$. As the noise level – and thus also the regularization parameter – approaches zero, the data fit term becomes dominant and the coefficient sequences less sparse. The solutions corresponding to the noise levels $\delta = 10^{-3}$ and $\delta = 10^{-1}$ can be seen in Figure 8.3, where the regularization parameters $\alpha = 10^{-4}$ and $\alpha = 3.872 \cdot 10^{-2}$, respectively, were chosen according to Morozov's discrepancy principle.

Additionally, we also plot the discrepancy functionals which we studied in great detail in Chapter 4 for a fixed noise level $\delta = 0.015$ and noisy data

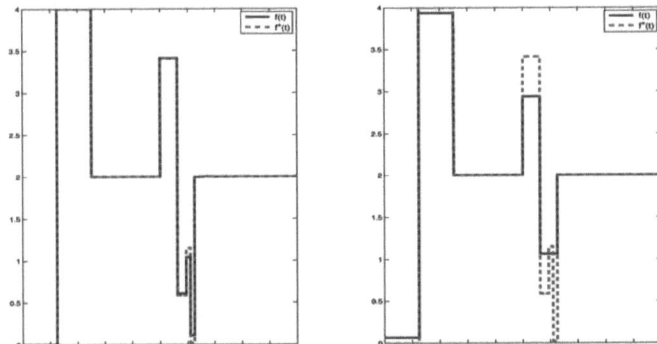

Figure 8.3: Regularized solutions with noise level $\delta = 0.001$ (left) and $\delta = 0.1$ (right) as solid lines together with the Ψ-minimizing solution $f^\dagger$ (dashed).

y^δ that was chosen such that we observe a jump in $g(\alpha)$ and $w(\alpha)$ near $\alpha = 0.1223$. In Figure 8.4 the functionals $g(\alpha)$ and $m(\alpha)$ are shown and $w(\alpha)$ is shown in Figure 8.5.

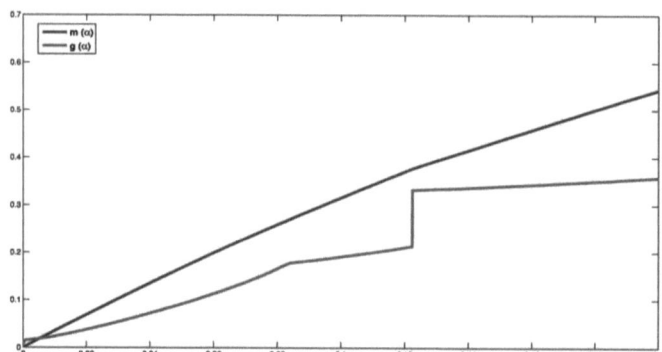

Figure 8.4: The discrepancy functionals $g(\alpha)$ and $m(\alpha)$ for $\delta = 0.015$ and fixed y^δ. Note that $m(\alpha)$ is continuous even though $g(\alpha)$ has a jump near $\alpha = 0.1223$.

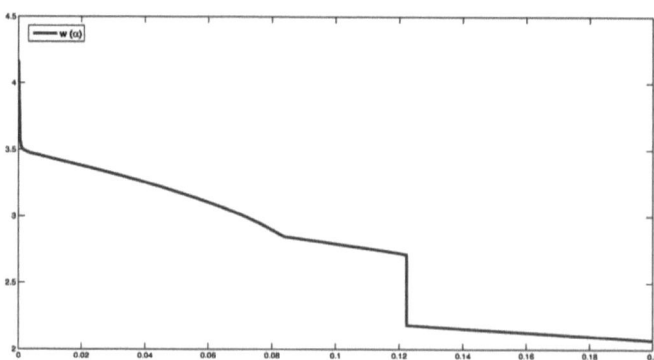

Figure 8.5: The discrepancy functional $w(\alpha)$ for $\delta = 0.015$ and fixed y^δ with jump near $\alpha = 0.1223$

Conclusion

In this work we have shown that the Tikhonov method with general convex penalty term $\Psi(x)$ and regularization parameter α chosen according to Morozov's discrepancy principle is a convergent regularization method for non-linear, ill-posed operator equations in Banach spaces.

Adapting results from [62] we showed weak convergence of the regularized solutions to the set of Ψ-minimzing solutions. For penalty terms fulfilling a generalized Kadec property (such as the ℓ_p norms of frame coefficients in a Hilbert space) this even implies strong convergence or convergence with respect to the penalty term (which may be even stronger than norm convergence).

We gave a sufficient conditions for the existence of a regularization parameter $\alpha = \alpha(\delta, y^\delta)$ that satisfies the discrepancy principle and for the so chosen parameter to satisfy $\alpha \to 0$ and $\delta^q/\alpha \to 0$ as the noise level δ tends to 0, where $q > 1$ may be chosen suitably for the problem under consideration.

If a searched-for solution $x^\dagger$ satisfies a generalized source condition and the operator under consideration a generalized non-linearity condition, both formulated as variational inequalities, then we found that the difference between the regularized solution obtained through our method and $x^\dagger$ when measured in the Bregman distance or a Taylor-type distance, goes to zero at a rate of δ^κ as $\delta \to 0$, where the parameter $\kappa \in (0, 1]$ allows for a relaxation of the classical source and non-linearity conditions, which are related to the case $\kappa = 1$.

Using another variational inequality (compare (6.20a)), which links the aforementioned Bregman- and Taylor distances to the Banach space norm, we could use the rates established for these distances to obtain a convergence rate $\mathcal{O}(\delta^{\kappa/r})$ in norm. Here the parameter $r \geq \kappa$ stems from the third variational inequality and even though such a constant r may be found from properties of the underlying (Banach) space and the penalty functional alone, it may be improved by additional knowledge about the true solution $x^\dagger$.

This behaviour could be observed when analyzing the situation of a solution which is known to be sparse in a Hilbert space setting, where the penalty term was chosen to be a weighted ℓ_p norm of frame coefficients with $1 \leq p \leq 2$. A rate with $r = 2$ can always be achieved for these choices, but using the sparsity assumption the third variational inequality could be shown to hold even for $r = p$, which yields convergence rates of up to linear order, $\mathcal{O}(\delta)$, in the limiting case $\kappa = p = 1$.

Finally, to substantiate our theoretical findings on a practical example we discussed the autoconvolution operator over a finite interval reconstructing sparse solutions with respect to a wavelet basis.

Bibliography

[1] Anzengruber S W and Ramlau R 2010 Morozov's discrepancy principle for Tikhonov-type functionals with nonlinear operators *Inverse Problems* 26(2) 025001 doi:10.1088/0266-5611/26/2/025001

[2] Anzengruber S W and Ramlau R 2011 Convergence rates for Morozov's discrepancy principle using variational inequalities *Inverse Problems* 27(10) 105007 doi:10.1088/0266-5611/27/10/105007

[3] Bakushinskii A B 1984 Remarks on choosing a regularization parameter using the quasi-optimality and ratio criterion *USSR Comput. Math. Phys.* 24 181–2

[4] Beck A and Teboulle M 2009 A Fast Iterative Shrinkage-Thresholding Algorithm for linear Inverse Problems *SIAM J. Imaging Sciences* 2(1) 183–202

[5] Boţ R I and Hofmann B 2010 An extension of the variational inequality approach for obtaining convergence rates in regularization of nonlinear ill-posed problems *Journal of Integral Equations and Applications* 22(3)

[6] Bonesky T 2009 Morozov's discrepancy principle and Tikhonov-type functionals *Inverse Problems* 25(1) 015015 doi:10.1088/0266-5611/25/1/015015

[7] Bonesky T, Kazimierski K S, Maass P, Schöpfer F and Schuster T 2008 Minimization of Tikhonov functionals in Banach spaces *Abstract and Applied Analysis* 2008 19p

[8] Bredies K, Lorenz D A and Maass P 2008 A generalized conditional gradient method and its connection to an iterative shrinkage method *Comp. Optim. Appl.* 42 173–193

[9] Burger M and Osher S 2004 Convergence rates of convex variational regularization *Inverse Problems* 20(5) 1411–21 doi:10.1088/0266-5611/20/5/005

[10] Burger M and Osher S 2011 A guide to the TV Zoo I: Models and analysis in *Level Set and PDE-Based Reconstruction Methods* to appear. (Springer)

[11] Cohen A 2003 *Numerical Analysis of Wavelet Methods* no. 32 in Studies in Mathematics and its Applications (Amsterdam: Elsevier Science B.V.)

[12] Daubechies I 1992 *Ten lectures on wavelets* no. 61 in CBMS-NSF Regional Conference Series in Applied Mathematics (Philadelphia, PA: SIAM)

[13] Daubechies I, Defrise M and DeMol C 2004 An iterative thresholding algorithm for linear inverse problems with a sparsity constraint *Commun. Pure Appl. Math.* 51 1413–41

[14] Daubechies I, Fornasier M and Loris I 2008 Accelerated Projected Gradient Method for Linear Inverse Problems with Sparsity Constraints *Journal of Fourier Analysis and Applications* 14(5-6) 764–792 doi: 10.1007/s00041-008-9039-8

[15] Daubechies I and Teschke G 2005 Variational image restoration by means of wavelets: Simultaneous decomposition, deblurring, and denoising *Appl. Comput. Harmon. Anal.* 19 1–16

[16] DeVore R A 1998 Nonlinear Approximation *Acta Numerica* pp. 51–150

[17] Engl H W, Hanke M and Neubauer A 1996 *Regularization of Inverse Problems* vol. 375 of *Mathematics and its Application* (Dordrecht: Kluwer Academic Publishers)

[18] Flemming J and Hofmann B 2010 A new approach to source conditions in regularization with general residual term *Numer. Funct. Anal. Optimiz.* 31 254–284

[19] Gfrerer H 1987 An a posteriori parameter choice for ordinary and iterated Tikhonov regularization of ill-posed problems leading to optimal convergence rates *Math. Comput.* 49 507–22

[20] Golub G H, Heat B and Wahba G 1979 Generalized cross-validation as a method for choosing a good ridge parameter *Technometrics* 21 215–23

[21] Gorenflo R and Hofmann B 1994 On autoconvolution and regularization *Inverse Problems* 10 353–373

[22] Grasmair M 2010 Generalized Bregman distances and convergence rates for non-convex regularization methods *Inverse Problems* 26(11) 115014

[23] Grasmair M 2010 Non-convex sparse regularisation *J. Math. Anal. Appl.* 365(1) 19–28

[24] Grasmair M, Haltmeier M and Scherzer O 2008 Sparse regularization with ℓ^q penalty term *Inverse Problems* 24(5) 1–13

[25] Grasmair M, Haltmeier M and Scherzer O 2009 The Residual Method for Regularizing Ill-posed Problems *Photoacoustic Imaging in Medicine and Biology Report* 14

[26] Grasmair M, Haltmeier M and Scherzer O 2011 Necessary and Sufficient Conditions for Linear Convergence of ℓ^1-Regularization *Comm. Pure Appl. Math.* 64(2) 161–182

[27] Griesse R and Lorenz D A 2008 A semismooth Newton method for Tikhonov functionals with sparsity constraints *Inverse Problems* 24(3) 035007

[28] Hadamard J 1902 Sur les problèmes aux derivées partielles et leur signification physique *Bull. Univ. Princeton* 33 49–52

[29] Hadamard J 1932 *Le probleme de Cauchy et les équations aux derivées partielles lineaires hyperboliques* (Paris: Hermann)

[30] Hansen P C 1992 Analysis of discrete ill-posed problems by means of the L-curve *SIAM Rev.* 34 561–80

[31] Hansen P C and O'Leary D P 1993 The use of the L-curve in the regularization of discrete ill-posed problems *SIAM J. Sci. Comput.* 14 1487–503

[32] Hofmann B, Düvelmeyer D and Krumbiegel K 2006 Approximate source conditions in Tikhonov regularization – new analytical results and some numerical studies *Mathematical Modelling and Analysis* 11(1) 41–56

[33] Hofmann B, Kaltenbacher B, Poeschl C and Scherzer O 2007 A convergence rates result for Tikhonov regularization in Banach spaces with non-smooth operators *Inverse Problems* 23(3) 987–1010

[34] Hofmann B and Scherzer O 1994 Factors influencing the ill-posedness of nonlinear problems *Inverse Problems* 10(6) 1277–97 doi:10.1088/0266-5611/10/6/007

[35] Hofmann B and Yamamoto M 2010 On the interplay of source conditions and variational inequalities for nonlinear ill-posed problems *Applicable Analysis* 89(11) 1705–1727 doi:10.1080/00036810903208148

[36] Ito K, Jin B and Zou J 2011 A new choice rule for regularization parameters in Tikhonov regularization *Applicable Analysis*

[37] Jin B and Zou J Iterative parameter choice via discrepancy principle *preprint*

[38] Justen L and Ramlau R 2009 A general framework for soft-shrinkage with applications to blind deconvolution and wavelet denoising *Appl. and Comp. Harmonic Anal.* 26(1) 43–63

[39] Kindermann S and Neubauer A 2008 On the convergence of the quasi-optimality criterion for (iterated) Tikhonov regularization *Inv. Prob. Imag.* 2(2) 291–299

[40] Kunisch K and Zou J 1998 Iterative choices of regularization parameters in linear inverse problems *Inverse Problems* 14(5) 1247–64

[41] Lindenstrauss J and Tzafiri L 1977 *Classical Banach Spaces I: Sequence Spaces* vol. 92 of *Ergebnisse der Mathematik und ihrer Grenzgebiete* (Berlin, Heidelberg, New York: Springer-Verlag)

[42] Lorenz D A 2008 Convergence rates and source conditions for Tikhonov regularization with sparsity constraints *Journal of Inverse and Ill-Posed Problems* 16 463–478

[43] Lorenz D A and Trede D 2009 Optimal convergence rates for Tikhonov regularization in Besov Scales *Journal of Inverse and Ill Posed Problems* 17(1) 69–76

[44] Loris I, Bertero M, DeMol C, Zanella R and Zanni L 2009 Accelerating gradient projection methods for ℓ_1-constrained signal recovery by steplength selection rules *Appl. Comput. Harmon. Anal.* 27 247–254

[45] Morozov V A 1984 *Methods for solving incorrectly posed problems* (New York: Springer-Verlag)

[46] Neubauer A 2010 Modified Tikhonov regularization for nonlinear ill-posed problems in Banach spaces *Journal of Integral equations and applications* 22(2) 341–51

[47] Ramlau R 2002 Morozov's discrepancy principle for Tikhonov regularization of non-linear operators *Journal for Numer. Funct. Anal. and Opt.* 23 147–172

[48] Ramlau R 2008 Regularization properties of Tikhonov regularization with sparsity constraints *Electron. Trans. Numer. Anal.* 30 54–74

[49] Ramlau R and Resmerita E 2010 Convergence rates for regularization with sparsity constraints *Electronic Transactions on Numerical Analysis* 37 87–104

[50] Ramlau R and Teschke G 2006 A Tikhonov-based projection iteration for non-linear ill-posed problems with sparsity constraints *Numer. Math.* 104(2) 177–203 doi:10.1007/s00211-006-0016-3

[51] Resmerita E 2005 Regularization of ill-posed problems in Banach spaces: convergence rates *Inverse Problems* 21 1303–1314

[52] Resmerita E and Scherzer O 2006 Error estimates for non-quadratic regularization and the relation to enhancement *Inverse Problems* 22 801–814

[53] Runst T and Sickel W 1996 *Sobolev spaces of Fractional Order, Nemytskij Operators, and Nonlinear Partial Differential Equations* vol. 3 of *De Gruyter Series in Nonlinear Analysis and Applications* (Berlin: Walter de Gruyter)

[54] Scherzer O 1993 The use of Morozov's discrepancy principle for Tikhonov regularization for solving non-linear ill-posed problems *SIAM J. Numer. Anal.* 30(6) 1796–1838

[55] Scherzer O, Engl H W and Kunisch K 1993 Optimal a posteriori parameter choice for Tikhonov regularization for solving nonlinear ill-posed problems *SIAM J. Numer. Anal.* 30 1796–838

[56] Scherzer O, Grasmair M, Grossauer H, Haltmeier M and Lenzen F 2009 *Variational Methods in Imaging* (New York: Springer-Verlag)

[57] Schock E 1985 Arbitrarily slow convergence, uniform convergence and superconvergence of Galerkin-like methods *IMA J. Numer. Anal.* 5(2) 153–160

[58] Tautenhahn U and Hämarik U 1999 The use of monotonicity for choosing the regularization parameter in ill-posed problems *Inverse Problems* 15 1487–505

[59] Tautenhahn U and Qi-nian J 2003 Tikhonov regularization and a posteriori rules for solving nonlinear ill posed problems *Inverse Problems* 19(1) 1–21

[60] Teschke G and Borries C 2010 Accelerated projected steepest descent method for nonlinear inverse problems with sparsity constraints *Inverse Problems* 26(2) 025007 doi:10.1088/0266-5611/26/2/025007

[61] Tikhonov A N and Arsenin V Y 1977 *Solutions of Ill-posed Problems* (Washington, DC: V. H. Winston & Sons)

[62] Tikhonov A N, Leonov A S and Yagola A G 1998 *Nonlinear Ill-posed Problems* (London: Chapman & Hall)

[63] Werner D 1995 *Funktionalanalysis* (Berlin, Heidelberg, New York: Springer-Verlag)

[64] Xie J and Zou J 2002 An improved model function method for choosing regularization parameters in linear inverse problems *Inverse Problems* 18(3) 631–43

[65] Zarzer C A 2009 On Tikhonov regularization with non-convex sparsity constraints *Inverse Problems* 25(2) 025006

[66] Zeidler E 1986 *Nonlinear Functional Analysis and its Applications I: Fixed Point Theorems* (New York: Springer-Verlag)

i want morebooks!

Buy your books fast and straightforward online - at one of world's fastest growing online book stores! Environmentally sound due to Print-on-Demand technologies.

Buy your books online at
www.get-morebooks.com

Kaufen Sie Ihre Bücher schnell und unkompliziert online – auf einer der am schnellsten wachsenden Buchhandelsplattformen weltweit! Dank Print-On-Demand umwelt- und ressourcenschonend produziert.

Bücher schneller online kaufen
www.morebooks.de

VDM Verlagsservicegesellschaft mbH
Heinrich-Böcking-Str. 6-8   Telefon: +49 681 3720 174   info@vdm-vsg.de
D - 66121 Saarbrücken   Telefax: +49 681 3720 1749   www.vdm-vsg.de

Printed by Books on Demand GmbH, Norderstedt / Germany